THE VEGETATION PROCESS

A HOLISTIC STUDY OF LONG-TERM ENERGETICS IN EAST BERINGIA

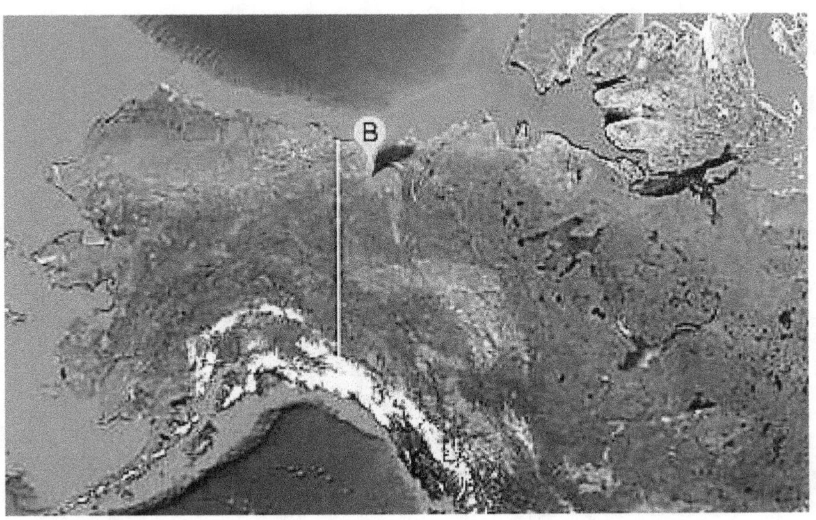

Google map of Alaska and the Yukon. Conterminous lighter hue on map, mainly to the West of Site B, represents the vegetation domain, East Beringia. Site B: the L.C. Cwynar core site (68.355787 N, 138.359613 W), the Book's principal data source.

What is inside the Book?

The focus is on the conceptualisation and actual measurement of potential energy oscillations in the vegetation's community assembly/disassembly process. The approach is quantum theoretical, but the basics are formulated in an ecological context. Numeric tables of results and graphs are included and explained. Regarding the required technical skills, many of the analytical tasks can be performed on a spreadsheet. Others require access to advanced software for derivatives and regression analysis.

To the memory of my Father
Officer of the Hungarian Panzer

THE VEGETATION PROCESS

A HOLISTIC STUDY OF LONG-TERM COMMUNITY ENERGETICS IN EAST BERINGIA

László Orlóci FRSC
Western University, London, Canada

SCADA Publishing - London – Canada

Refer to this monograph:

Orlóci, L. 2014. The vegetation process. A holistic study of long-term community energetics in East Beringia. SCADA Publishing, Canada. Online Edition: https://createspace.com/4760258

Look for these:

Orlóci, L. 2013. Quantum analysis of primary succession. The energy structure of a vegetation chronosere in Hawai'i Volcanoes National Park. SCADA Publishing, Canada. Online Edition: https://createspace.com/4452597

Orlóci, L. 2013. Quantum Ecology. Energy structure and its analysis. SCADA Publishing, Canada. Online Edition: https://createspace.com/4406077

Orlóci, L. 2013. On the Energy Structure of Natural vegetation. In search for community governance rules. SCADA Publishing, Canada. Enlarged Online Edition: https://createspace.com/4153484

Orlóci, L. 2012. Self-organisation and Mediated Transience in Plant Communities. SCADA Publishing, Canada. Enlarged Online Edition: https://createspace.com/3585127

Orlóci, L. 2012. Statistical Ecology. The quantitative exploration of nature to reveal the unexpected. SCADA Publishing, Canada. Online Edition: https://createspace.com/3476529

Orlóci, L. 2012. Statistical multiscaling in dynamic ecology. Probing the long-term vegetation process for patterns of parameter oscillations. SCADA Publishing, Canada. Online Edition: https://createspace.com/3830594

Orlóci. L. 2011. Problem flexible computing in statistical ecology. SCADA Publishing, Canada. Online Edition: https://createspace.com/3574792

Title ID: 4760258
ISBN-13: 978-1499142068
ISBN-10: 1499142064
V 2014-06-27

Find further information at URL
https://sites.google.com/site/statisticalecology/

THE VEGETATION PROCESS

Contents

Preface

Process, as the Book uses this term, implies simultaneous execution of two fundamental functions in continuity. One creates complexity, the other reduces it. Ecologists refer to these as community assembly and disassembly. The process requires energy input which determines the potential energy state of the community. This is measurably true in terms of Max Planck's *energy-based entropy.* We find potential energy increasing when new species (taxa, community elements) are added or others proliferate in the community, and decreasing when species drop out or their performance declines. Data analysis which targets energy-based entropy is identified in the Book as *Quantum analysis.*

Quantum analysis identifies the data analytical component of a holistic paradigm for the study of ecological energetics. Holistic implies that energetics is studied at the community level. I describe the basics of ecological quantum analysis in a separate essay.[1] I give a brief summary later in the text and include new extensions, both conceptual and technical.

I consider it important that students learn early what kind of energetics is on the analytical palette at this time and refrain

[1] Orlóci, L. 2013. Quantum Ecology. Energy structure and its analysis. SCADA Publishing, Canada. Online Edition: https://createspace.com/4406077

from thinking tangentially following the comforting lure of what they already know about *calorific flow*[2] when in fact quantum analysis is about *potential energy*.

Quantum analysis revolves about entropy, yet it has only far cousins in the information theoretical world of A. Rényi's generalised entropy and information or in S. Kullback's information analysis centred on his minimum discrimination information statistic. Indeed, quantum analysis was born to fill the void in energetics concerned with potential energy. Max Planck gave as the parameter $E=\ln\frac{1}{P}$ for potential energy in complex systems of the resonator complex type.

Beringia is Hultén's (1937) term. He used it for that vast, climatically arid arctic zone which takes up a good part of Alaska, extends westward into Siberia and eastward into the Yukon Territory. Exactly for the climatic aridity, Beringia escaped the direct effects of glacial advances and glacial retreats in the Quaternary.

Quaternary is the scientific name for the geological period which began approximately 2.6 million years ago. The period had seen massive glacial advances and retreats on the Northern Hemisphere in cycles of rigid regularity. Milankovitch (1947) explains the triggering mechanisms in his model of the Earth's orbital movements.[3] The Milankovitch model has much to tell to society. The main message is that interglacial periods - such as the present, during which our civilisation evolved - are indeed preordained events. Interglacial periods have limited time spans. Being followed by a new Ice Age is not just a cer-

[2] ... for which H.T. Odum (1871) erected lasting epitaph in late 19th Century ecology.

[3] http://en.wikipedia.org/wiki/Milankovitch_cycles

tainty but already overdue. It is so ironic that much before it happens, the ongoing, fast developing and potentially most destructive global warming event may have run its full course.

With no erosive force of the magnitude and generality of the advancing glacial ice, the land surface in Beringia retained the conditions favourable for deep, plant particle bearing sediments to be preserved. One of the deepest is in northern Yukon's Hanging Lake. The bottom layer in the sediment is carbon-dated to about 42 thousand years. This period is the depth of Book's vista onto the *Late-Quaternary* epoch.

The Book takes aim at the dynamics of the vegetation cover's potential energy state[4] through two major climatic cycles of the past 42k years in Beringia. One characterised by cooling and the other by warming as the dominant trend. I refer to my subject in this dynamics by the term *holistic vegetation energetics*. With this I intend to emphasise that what I do in the analysis has very little to do with the so much worked notion of "calorific flow", but everything as I already mentioned with notion of "potential energy". I explain this in the sequel. Just that much more at this point that the quantum ecological techniques are adapted at this time to Cwynar's (1982) Hanging Lake plant particle chronosere and presented as such.

It is important to keep the definitions straight and particularly to re-emphasise that we are not doing "calorific flow" which most of us learnt to associate with the well-drawn, richly decorated, low-information charts inundated with arrows pointing in all directions. I know that these charts are just a face, not the science of a complex reductionist approach.

[4] This will be explained, but the reader need not wait. Any introductory text in Physics should have potential energy defined in more details than I have room for in this Essay.

I missed in studies of energetics a holistic definition of the vegetation's energy structure. I needed one for the study of long-term dynamic patterns in energetics for the effective analysis of plant particle chronoseres. Max Planck's 1901 quantum theoretical paper came to my rescue. His focus was on potential energy defined in the manner of energy-based entropy. This is the logarithm of the inverse of probability. The probability refers to any resonator complex being by pure chance exactly such as one actually observed.

I consider my dialect of quantum analysis a "high level" diversity analysis of complex systems. Hence is my characterization "holistic". Diversity analysis on this level is unlike the usual type in ecology which penetrates all the way down to the resonator probabilities in the systems. I associate the latter from my student years with names including Shannon (1948), Rényi (1961) and Kullback (1959). They created new world of thought for me; E.C. Shannon by his use of entropy to capture information ecologists call diversity, A. Rényi by his brilliant generalization of entropy and information to different orders which showed ecologists how their diversity indices are generically connected, and S. Kullback by introducing me to information based Statistics, a paradigm for probabilistic data analysis without sums of squares and products.

I thank Eng. Márta Mihály and Prof. Mathew Mukkattu for suggesting needed revisions and corrections after reading an earlier version of the manuscript.

London
April 12, 2014

Why Beringia?

My involvement in the region's ecological research reaches back to the late 1970s. At that time I designed and directed with my good friend Walter Stanek the vegetation survey on Foothill's gas pipeline right of way along the Alaska Highway's entire Yukon section (Orlóci and Stanek 1980)[5]. My closer interest in the vegetation ecology of the Yukon Territory was rekindled some years later after reading L.C. Cwynar's (1982) paper on his long, plant particle chronosere from Hanging Lake.[6] I used this chronosere in the development and testing of multiscale trajectory analysis (Orlóci 2009 and references therein)[7].

I reach back to L.C. Cwynar's data now the second time. I do this with salutation and expression of gratitude for the work done and for making the entire data set available in the public domain with minimum restrictions.

What makes the Hanging Lake site so unique to me is its location in the eastern most corner of the Beringian tundra, along

---

[5] László Orlóci and Walter Stanek 1980 – visit https://sites.google.com/site/statisticalecology/ and click item 41 in the list of selected references.

[6] http://hurricane.ncdc.noaa.gov/pls/paleox/f?p=519:1:0::::P1_STUDY_ID:7483
While this data set is in the public domain, it would not be in existence without perseverance through all difficulties of field work on the arctic Tundra by the data set's authority, L.C. Cwynar (1982).

[7] László Orlóci 2009 -- http://link.springer.com/article/10.1007%2Fs11515-009-0012-y

the Beaufort Sea, where current climate warming has already demonstrated with force its perturbing effects.[8]

The coupling of historic compositional transitions in the vegetation with the latitudinal effect - which exaggerates the climate warming rate in the Arctic tundra many times over the global average – was the topic in earlier papers on climate warming (Orlóci 1994, 2008).[9] With the L.C. Cwynar chronosere in hand, I can take a step forward towards laying open governance principles which drive the long-term vegetation process within strict limits set by the climate. At this time I have vegetation transitions in the focus, but I approach the process as a potential energy structural phenomenon.

---

[8] Early signs and prospects in the North: http://ny.water.usgs.gov/projects/climate/YukonClimate.pdf

[9] László Orlóci 1994,2008 – visit https://sites.google.com/site/statisticalecology/ and click item 15 in the list of selected references; also http://www.akademiai.com/content/l246056112n4/?k=Orloci

L.C. Cwynar's chronosere

The sampled site is Hanging Lake in arctic Yukon (68.355787 N, 138.359613 W, marker B in Figure 1). The Lake is located in ragged terrain along a sand stone ridge approximately 83 aerial kilometres due south from the nearest coast line of the Beaufort Sea.

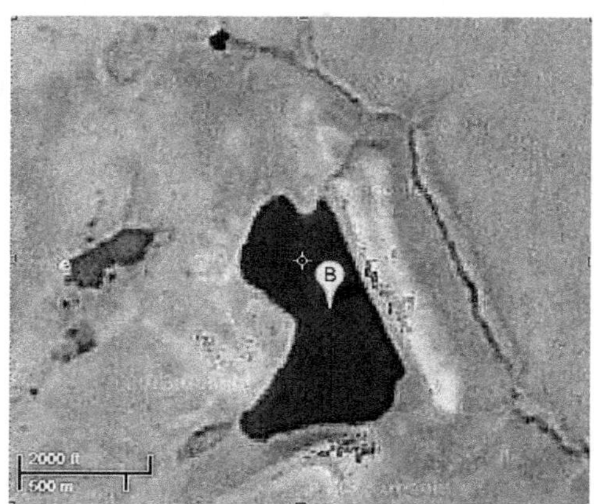

Figure 1. L.C. Cwynar's sampling site, Hanging Lake, Yukon Territory. Note the sharp straight lines, indicative of terrain which escaped the effects of glacial ice.

I took the entire Cwynar record set[10] as I found it from the NOAA data base. The records detail particle counts for 89 palynomorph taxa in 133 paleorelevés. One paleorelevé is marked by the horizontal line in Figure 2. Note the reduced set of taxa.

L.C. Cwynar's paper gives details concerning the current vegetation at the site in which he identifies 47 species of vascular plants, 12 bryophytes, and 18 lichens of several tundra types.

---

[10] Cwynar 1982 – http://hurricane.ncdc.noaa.gov/pls/paleox/f?p=519:1:0::::P1_STUDY_ID:7483 Global Pollen Database (2000) www.ncdc.noaa.gov/paleo/ftp-pollen.html

The types are named after the vegetation's physiognomy (tussock, heath) or the quality of the substrate (shale felt fields, and sandstone slope).

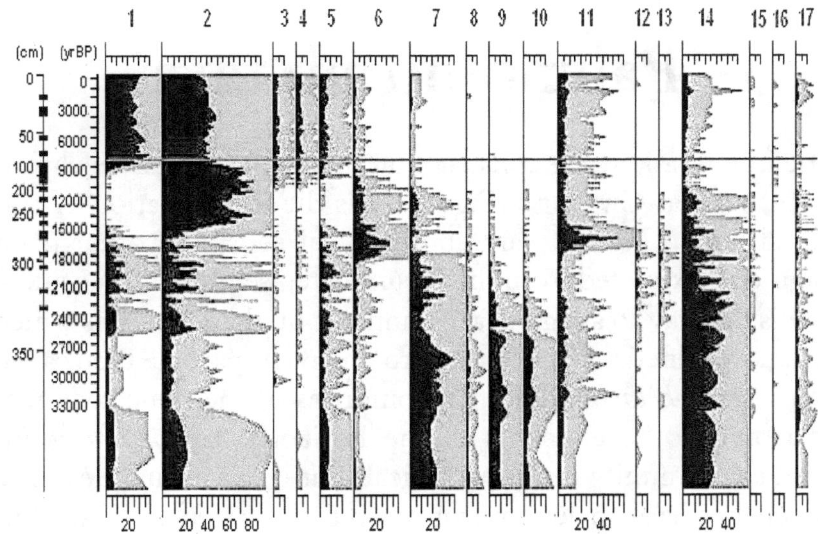

Figure 2. Plant particle identities and distribution by sediment horizon in L.C. Cwynar's chronosere. Seventeen shortlisted taxa are included. Graph and accompanied data are available from the NOAA data base (see URL in footnote 10). Short listed taxa: 1 Alnus, 2 Betula, 3 Ericaceae, 4 Ericales,[11] 5 Picea, 6 Salix, 7 Artemisia, 8 Asteraceae/Asteroideae, 9 Brassicaceae, 10 Chenopodiacea/Amaranthaceae, 11 Cyperaceae, 12 Fabaceae, 13 Plantago canescens, 14 Poaceae, 15 Rosaceae. No 16 in the graph identifies "other trees and shrubs" and 17 "other herbs". Bottom scale: pollen counts %. Dark shading: original scale. Light shading: 5x original scale. Black markings on depth scale: carbon dated horizons. Bottom layer in sediment dated 41138 year old. Horizontal line at 8299 yr. BP: a record set, the *paleorelevé*.

It is interesting to observe the sediment age. From this we can readily infer an unglaciated status for more than 40 millennia.

[11] Not an error. Ericaceae and Ericales are different palynomorph taxa. See next chapter.

Palynomorph taxa

The initial task in the construction of any plant particle chronosere, such as Cwynar's (Figure 1), is the identification of fossilised pollen grains, plant spores, algal cells, and other plant fragments extracted from the accumulated sediment. Identification is the way to assign a large number of plant particles which carry a specified set of key traits to one plant particle type called *palynomorph taxon*. The palynomorphs of the same sediment horizon are the elements of the horizon's *paleocommunity*. Such a community is not biological. But it has virtual existence as a chance based palynomorph assemblage, a sample of the limitless metacommunity whose natural concentric belts have diminishing chance of being represented in the sample in proportion to their availability from the site of the sediment core. Air currents, flowing water, gravity and members of the fauna with which the plant particles come into contact are the selecting agents. The vector description of a paleocommunity is the *paleorelevé*.

It is important to realise that the usual state of sediment-born plant particles is rather poor and plant identification based on the usual taxonomic criteria of the modern phylogenetic systems bounds to meet with only limited success. For example: we already encountered the palynomorph taxon Ericaceae and the other Ericales. The presence of both in Cwynar' list implies that he could not take the identification to lower levels of plant systematics in concrete cases of some particles. He accepted Ericales and Ericaceae as distinct palynomorph taxa. Therefore, names Ericales and Ericaceae are used as labels of two valid palynomorph taxa.

Obviously the use of a recognised taxonomic nomenclature should not mislead anyone to thinking of phylogenetic homogeneity within the palynomorph taxa. These are not species, not genera, not families, and not anything of that sort. Far from it, they are simply distinct palynomorph types representing distinct populations of plant particles. In such populations the mixed inheritance of the parental sources is the rule, not the exception.

Quantum ecology

Preliminaries

I do not intend to re-explain the basic theory or develop new, step-by-step training exercises. There is much on those in my earlier essays (see page 3). I suggest Cwynar's (1982) original paper for reading and similarly Max Planck's (1901) seminal paper for study for students before progressing to the next section.

Potential energy

We learned in introductory physics that an object whose mass is M = 1 kg, lifted up to height h in meters (m) against gravitational acceleration g per second has potential energy level $\varepsilon = gh$ $mkgs^{-2}$. Let the object be one of n in a set with average mass M and total mass T=nM. The potential energy level in the set is then $E = T\varepsilon$ $mkgs^{-2}$.

In the exercise above we counted the number of objects to find a value for n and summed their mass values to find the total T. These specify the data type on which we perform ecological quantum analysis. Interpreted for just one paleorelevé, we have:

n number of palynomorph taxa in the relevé, up to 93 in Cwynar's chronosere,
X – one taxon particle count M, i.e. the number of energy units, and
T = nX

We designate the total number of paleorelevés in the following by letter s (133 in the Cwynar chronosere). In these terms the value of ε is the potential energy level of one taxon in the relevé. Accordingly, we can write for the state of potential energy the paleorelevé $E=n\varepsilon$, but not in $mkgs^{-2}$ units. If not in such units, then in what kind of units?

Since the choice of unit to express energy is completely arbitrary[12] we are free to opt for any which we find useful for our purposes. I choose the natural unit and designate it by *nat.* This choice takes us by way of Max Planck's (1901) reasoning to his energy-based entropy function:

$$E=nH=-\ln P=\ln C=(T+n)\ln(T+n)-T\ln T-n\ln n$$

in nats. Additional definitions, $C=\frac{(T+n-1)!}{T!(n-1)!}\approx\frac{(T+n)^{T+n}}{T^T n^n}$

by Stirling's approximation, and further, $P=\frac{1}{C}$ and $H=\frac{nH}{n}$ nats

Clearly, Max Planck uses energy based entropy nH via a probability P to potential energy. In fact P is the probability that an n-valued complex exactly the same as the paleorelevé we actually observe can occur under the total rule of chance.

Numerical example

We apply what we already defined to the paleorelevé of carbon year 24136 BP (see horizontal line in Figure 2). This relevé has 23 palynomorph taxa (n) and total plant particle counts rounded to 384 (T):

[12] Richard Feynman, The Feynman Lectures on Physics (1964) Volume I, 4-1 – http://en.wikipedia.org/wiki/Energy
http://en.wikipedia.org/wiki/The_Feynman_Lectures_on_Physics

Taxa	Particle count		
54 Poaceae	126	82 Pediastrum	3
5 Betula	61	1 Alnus incana	2
23 Artemisia	37	57 Potentilla	2
2 Alnus viridis	35	51 Phlox	2
12 Picea	27	83 Sphagnum	1
32 Cyperaceae	25	76 Lycopodiaceae	1
27 Brassicaceae	22	24 Asteraceae subf.	1
30 Chenopod/Amar.	13	26 Astragalus	1
80 Polypodiaceae	7	37 Fabaceae undiff.	1
15 Salix	6	25 Asteraceae	1
52 Plantago canes	6	64 Saxifraga cernua	1
71 Botrychium	3		

Numerical label preceding the taxon's name indicates position (rank) in the original data set. Names are taken directly from the original record as found. Figure 3 has the relevé's graph.

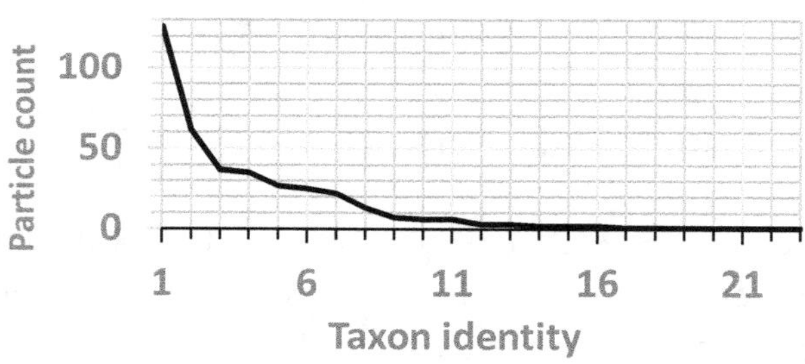

Figure 3. Graph of the paleore-levé crossing at year 24136 BP in Figure 2. The sediment debt at this point is close to 100 cm. The relevé has 23 palynomorph taxa. Relevé composition indicates with clarity the zonal presence of the cold arid steppe in the region at 24k BP. Quantum theoretical quantities:

n	T	ECPR	nH	H	P1	C1	P2	C2
23	384	17	88.4239	3.8445	3.96E-39	2.52E+38	0.0214	46.7360

Legend to symbols: n – number of taxa, T – paleorelevé total rounded to nearest integer, ECPR – number of energy units (quanta) of one taxon, nH – the relevé's potential energy level in energy-based entropy units, H – potential energy level of one taxon, $C1=P1^{-1}= \exp(nH)$, $P2=\exp(-H)$.

Further explanations will be provided in the sequel. There are questions the readers may want me to answer before we move to the next section: What justifies using measurements as energy unit counts? What justifies the use of nH or H as proxy for the potential energy state of a vegetation community? The answers should point back to two of Max Planck's quantum theoretical principles:

1. Energy is transmitted in discrete units, the quanta. Therefore energy units are countable.

2. nH (or H) and energy are interchangeable parameters of *resonator complexes.* The relevé is such a complex and the taxa are resonators.

Further regarding the terminology, designation of the paleorelevé as a complex and the taxa as its resonators is provisional. There will be cases where we have to redefine complex and resonator. Functional types are cases in point. We consider the topic further in a different section.

More on energy-based entropy

Returning to the equations, we see the dependence of nH and H on richness n modified by abundance T. If we assume that an observed relevé's composition is the outcome of a completely random process - an impossible state of natural vegetation, but still a very useful concept pf statistical theory – the probability that observed relevé is $P= C^{-1}$.

A further point to be remembered is this: when H is given, a probability is defined, $P=e^{-H}$. In a statistical test H is considered statistically unique (we may use the term *significant*) when it is at least 3 nats or when e^{-H} is not higher than 0.05.

Why quantum analysis?

The reader's next question may very well be "why should anyone in ecology be interested in quantum analysis? My answer has several folds:

1. I have not found yet another way to perform a holistic analysis on the energy structure of vegetation communities. The holistic approach[13] ties me down to potential energy and directs me to Max Planck's energy based entropy function to measure energy levels.

2. Quantum analysis can isolate footprints left by historic processes in the potential energy structure of the vegetation. This is a main objective when considering long chronoseres, such as Cwynar's.

3. Quantum analysis offers facilities for measurements of the homogeneity and stability/instability of the plant community's energy structure and facilities the detection of superposition (emergent energy) in chronoseres.

Potential energy and diversity

I mentioned earlier that quantum analysis is complex level diversity analysis. What distinguishes such a diversity analysis from other entropy based diversity analyses is energy-based entropy nH. Energy-based entropy is neither Rényi's (1961) gen-

---

[13] Thoughts expressed on holism and reductionism by Gleick (1987), Çambel (1993) and van Hulst (2000) - among others whom I quote in my essay "From Order to Causes" (Orlóci 2000) – are close to my views regarding the study scenario concerning complex systems.

eralised entropy $H_a = \frac{1}{1-a} \ln \sum_{i=1}^{n} p_i^a$ nor Rényi's generalised information $I_\alpha = \frac{1}{\alpha-1} \sum_{i=1}^{n} \frac{p_i^\alpha}{q_i^{\alpha-1}}$ (see Orlóci 2006, 2012). Rényi's equations require p and q values. For p we have to reach back to the basic elements in the T-totalled data vector **X** and define p_i as $\frac{X_i}{T}$. The q values are expectations from hypothesis or theory.

Note, in quantum analysis P is a combinatorial function of T and n. The reader should note further that the similarity of quantum theory's $nH = \ln C$ and Brillouin's information function $I = \log_2 C$ is superficial.[14] Brillouin's C is defined by $\frac{N!}{f_1!\, f_2! \ldots f_s!}$ in which f_i can be an element of my X from which p_i of Rényi's equation is calculated, so that $p_i = \frac{f_i}{N}$.

Yet another important point, Brillouin's information I is not a divergence measure, therefore it is not the same as Rényi's I or Kullback's minimum discrimination information.

Once more, in Planck's nH, Rényi's p and q parameters do not receive direct consideration. The parameterisation of the energy equation relies entirely on T and n. This should really reveal to the student the very essence of where my adaptation of

[14] Note: $\log_2 2$ is one bit and the maximum I is $\log_2 s$ bits. Brillouin shows that when the f_i are large, say 100 each or greater, then $\frac{I/n}{\log_2 e}$ will come close in value to Shannon's (1948) entropy function. In this, *e* is the natural base (2.718281828).

quantum analysis draws information from in ecological diversity analysis. Clearly this information comes from a high level of organization, above the resonator level on the level of the complex itself.

Structural instability

My definition of the observed complex's structure instability is based on P. For this we do not set P's lower limit to zero, we let its value approach it. At $P = 0$ there is no complex in existence. We do not put P to 1 either. $P=1$ would imply that the natural complex can exist in a singular state for perpetuity.

We take the case where $0 < P < 1$. In other words the complex that merits our further interest must have at least two states a and b such that $P_a=\frac{1}{C_a}$ and $P_b=1-P_a$. This implies that the complex can exist simultaneously in two states with probability $P_{ab}=P_aP_b$.

We can derive based on the probabilities three scalars such as in the expression $w_a+w_b+w_{ab}=1$. For example, if $P_a=0.8$ then $P_b=0.2$ then we have,

$w_a=P_a^2=0.64$, $w_b=P_b^2=0.04$ and $w_{ab}=2P_{ab}=1-0.64-0.04=0.32$

Obviously, the closer w_{ab}, the instability scalar comes in value to 0.5 the easier is for the state of the complex to flip from one state into the other state. This w_{ab} has values within 0 and 0.5.

Superposition

The superposition principle of complex systems has close allies in the ecological notion of emergence. This is expressed in the frequently quoted textbook phrase "the whole is more than the sum of its parts". But, a complex system to be more than the sum of its parts has to have parts hidden from direct observation. We refer to these in various ways as ghost, intrinsic, emergent in combination with states, structures, traits, and so forth.

The theoretical existence of ghost states is explicitly stated in the superposition principle, which I paraphrase: the complex exists in all its particular theoretically possible states, but when measured or observed, the result describes only one of the possible states. I have good examples from phtosociology to illustrate what this is really about.

Borhidi (1961) examined climatic records of the Tisza Plains and found that the Plains had the Pontic steppe climate[15] in 20 out of 100 years on record. The Zólyomi, Kéri and Horváth (1997) spectrum gives these details:

Climate year type	Number of years
Subatlantic-Alpean	9
Atlantic-Submediterranean	24
Euro-Continental	41
Pontic Submediterranean	20
Other	6
Total	100

Suppose we identify the climate year as the Pontic steppe type. This is only one of the five climate types. The other four must also exist because their probability of occurrence is not zero. These are ghost states of the complex specified by the Borhidi-Zólyomi spectrum.

The question of what is the probability that a spectrum, call it a complex, could occur by pure chance in every detail exactly the

[15] The Pontic steppe climate happens to be the dominant climate type of the Black Sea's west coastal region according to Köppen's criteria.

same as the Borhidi-Zólyomi spectrum. In theory the number of possible complexes each having 5 resonators and a total of 100 energy units counted is

$$C \approx \frac{(T+n)^{T+n}}{T^T n^n} = \frac{105^{105}}{100^{100} 5^5} = 537064419$$

The one complex probability is P=1/C=1.86197402885481E-09. The observed complex is real, the other 537064418 are ghost complexes.

Given a value C, we can measure the potential energy state of the Borhidi-Zólyomi spectrum:

Number resonators n	5
Potential energy state in complex nH=	20.10162861
P=exp(-nh)	1.86197E-09
C=1/p	537064419
H=nH/n	4.020326
P=exp(-H)	0.017947
1-P	0.982053
H=exp(-1+P)	0.374541
$w_{ab}=1-P^2-(1-P)^2$	0.035250
Stability= 0.5-w_{ab}	0.464750

The first of the last two items is the potential energy-based instability coefficient. Its value is 0.035250 or a round 7%. The last is the stability coefficient. We conclude the complex's potential energy-based stability is very high 93%. What does this mean?

We should not forget that we are dealing with a complex whose potential energy structure carries the footprints of chance effects. In nature this can never be 0 nor 1 in proportional terms, but somewhere in between. The extremely low probability 1.86197E-09 makes us to regard the Borhidi-Zólyomi spectrum is a very unique event in the sense that it would be unusual to expect it under the theoretical condition we call total rule of chance. We conclude it is not a chance event in Nature and declare statistical significance. The stability analysis fully supports this conclusion. A complex of such a high stability has very little chance to flip in and out of potential energy states under random effects.

What can we distil from what we so far presented about superposition? -

1. When $P=\frac{1}{C}$ is greater than zero or less than 1, there exists in theory at least one ghost state whose probability is at most 1-P.

2. The chance of a complex with high stability flipping out of the observed potential energy state into one of the n-valued T-totalled ghost states by chance is very small.

3. Small probability means uniqueness and high stability implies resistance to chance effects.

All can change, of course, should the forcing environment change into another kind.

We shall consider the notion of superposition in another context when we examine the emergent potential energy cloud at pivotal paleorelevés, at which parts of a chronosere are joined.

What to expect?

On close observation of what ecologists do. we find that the aim is to probe Nature for the unexpected or verify the expected Using remote sensing as my example the image may be probed for patterns unspecified or subjects completely circumscribed. The quantum analysis serves both objectives in holistic studies.

In the present case quantum analysis illuminates the vegetation's long-term energetics at the level of palynomorph assemblages (paleocommunities) to reveal long-term cyclic patterns in potential energy oscillations, to detect and isolate the footprints of climatic events in the vegetation's potential energy structure, to measure the potential energy structure's instability level, to test superposition phenomena which as ghost clouds shadow pivotal vegetation states, and to measures the state of potential energy as a Newtonian force.

The nature of the analysed medium, the plant particle chronosere, has sever constraints on doing other interesting analysis, such as probing the vegetation's potential energy structure for footprints of the phylogenetic signal and its isolation from current and past environmental mediation. The reason for the constraint is the lack of unique evolutionary systematic status of the palynomorph taxa. The reader finds more on this topic in recent papers referenced on page 6 and in the Bibliography.

The reader can expect to see chance-effect management in similar ways as in any well-conceived statistical analysis: the observed parameter states' expectations under pure chance effects serve as convenient threshold states to which the actually observed values are compared in tests of significance.

Proxy climatic chronoseres

This section is concerned with causal conditions which can trigger global transitions in the vegetation's composition and potential energy levels on scales for which the Cwynar plant particle chronosere is Beringia's witness. Those conditions are in fact linked with the Late Quaternary global cooling and warming cycles (Figure 4A).

The following analysis examines the oscillograms of two long-term temperature chronoseres, one from the Vostok research site in Antarctica[16] and the other from the Renland research site on Greenland[17]. The intention of the analysis is to extract general trends in the manner of shape functions, to isolate the residuals, and to write differential equations. We need these to evaluate the proxy temperature chronoseres' relevance and to trace the temperature effect in the potential energy state of the vegetation.

Temperature chronoseres

My main source for the temperature cycles is Vostok. I use Renland in a limited way. The amount of deuterium ^{2}H in the ice is the proxy for the Vostok temperatures and the amount of ^{18}O and ^{16}O isotopes in the air bubbles trapped in the ice are the

---

[16] World Data Centre for Paleoclimatology (2002): www.ngdc.noaa.gov /paleo/icecore/antarctica/vostok

[17] ftp://ftp.ncdc.noaa.gov/pub/data/paleo/icecore/greenland/vinther2008renland-agassiz.txt

proxy for the Renland temperature values. I refer for details to Petit at al. (1999, 2001) regarding technical details for Vostok and Vinther et al. (2008, Hansson 1994) for Renland[18,19]. The temperature differences oscillograms are displayed by Figure 4A.

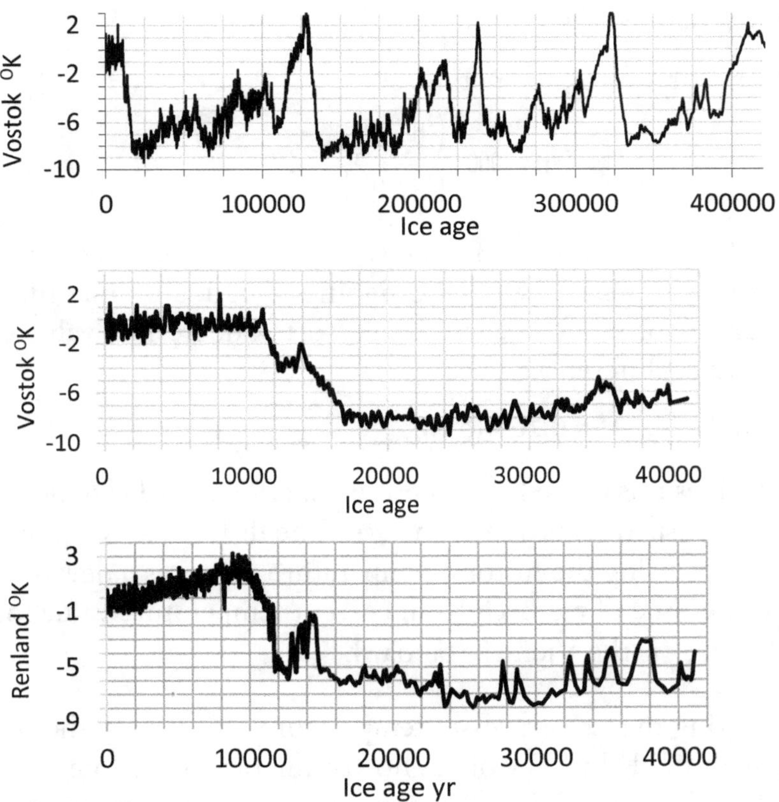

[18] ftp://ftp.ncdc.noaa.gov/pub/data/paleo/icecore/greenland/vinther2008renland-ag-assiz.txt

[19] http://www.tellusb.net/index.php/tellusb/article/view/15813

Figure 4A. Vostok and Renland temperature differences oscillograms.[20] *Top:* Vostok chronosere, 3311 points. *Middle:* Reduced length Vostok chronosere, 637 points. *Bottom:* Renland chronosere, 821 points. Time period covered: close to 42kyr, the time span of Cwynar's plant particle chronosere.

The following is intended for general information which I base on the quoted works:

1. Proxy for Vostok temperature:

$$\delta\,^2H = 1000\left[\frac{\left(^2H/^1H\right)_{sample}}{\left(^2H/^1H\right)_{vienstd}} - 1\right]‰$$

In this ^2H is the amount of deuterium and ^1H is the amount of ordinary hydrogen. The letter δ signifies that deuterium differences are involved. To get from the Vostok deuterium value to temperature differences I use

$$\pm t = 0.165[(\delta^2H)_{samplt} - \delta^2H_{current}]\ ^OK$$

Where does this come from? Petit et al. (1999, 2001) applied it whose published record set I analyse. The difference is equivalent to expressing the historic temperature values as a deviation at the time of research from current temperature value for which $\delta\,^2H_{current}$ is the Vienna Standard.

Considering that ±t expresses temperature differences, the replacement of ^OK by ^OC violates no arithmetic rules. Note that $1^OK=1^OC$. The coefficient 0.165 seems to be consistent with Schweingruber's (1996) regression equation,

---

[20] The alternative use of ^OC or ^OK in the literature on the same graph should not be confusing. Both are correct since temperature differences are expressed.

$$t = \frac{D - 129.3}{5.666} = 0.176\ D - 7.734\ ^{O}K$$

Schweingruber extracted ^2H from the cellulose of tree rings. His equation specifies temperature change on a geographic gradient from Yukon to Florida.

2. Temperature base for Renland is similar in form to $\delta^2H_{current}$; we take a difference from the standard and divide by the standard:

$$\delta^{18}O = 1000\left[\frac{\left(^{18}O/^{16}O\right)_{sample}}{\left(^{18}O/^{16}O\right)_{vienstd}} - 1\right]‰$$

In this the ratio $(^{18}O/^{16}O)_{vienstd}$ is the mean amount in sea water in accordance with the Vienna Standard. To get from the $\delta^{18}O$‰ value to a temperature scale I used

$$\pm t = \frac{\delta^{18}O_{sample} - \delta^{18}O_{current}}{\alpha}\ ^{O}K$$

In this α=0.67 ‰ corresponds to 1 ^{O}K temperature rise. Cuffey et al. (1995) recommended α=0.33.These are only two of many transformation coefficients in circulation. This warns us not to take the t values for more than praxis and to focus on the shape of the chronosere and not so much on the proxy temperature dimension.

At this point I can anticipate question that the reader probably wish to put to me:

Do I plan to use both temperature chronoseres? If I have to choose, which one I prefer for the Beringian Tundra? Is the arctic location of Cwynar's chronosere mandating the use of the Renland chronosere? The two chronoseres obviously differ, but the differences are more related to scale than shape (Figure

4A1). How do I isolate shape from size? On which I will consider to weigh in more heavily when making a choice?

I provide answers directly or implicitly by example but before that I want to get a complete analysis done on the Vostok and Renland chronoseres.

Finding functions to capture shape

The raw graphs such as Figure 4A1 are difficult to use. We can impart more generality to them if we separate the trend from random oscillations. The difference of the two is a set of residuals. To do this we chose for the graphs a mathematical equations. For the shorter chronoseres I suggest a simple yet very powerful equation which I call *shape function*,

$$f(x)=\frac{a + cx + ex^2}{1 + bx + dx^2}$$

The function $f(x)$ is parameterised in regression analysis. Parameterisation means finding real values a, b, c, d, e. Regression results are presented in Figure 4B and in the numeric tables following the figure.[21]

We need yet another set of functions for which we placed graphs in the figures. These are the 1st and 2nd order differential equations[22] for $f(x)=\frac{a + cx + ex^2}{1 + bx + dx^2}$:

---

[21] I used TableCurve 2D v4 in the analysis. Visit http://systat.co.kr/products/TableCurve2D/resources/help.html for general information and technical terms. "Statistical ecology" (Orlóci 2012) is a source for concepts and technical details used in ecology.

[22] The interested user not versed in calculus need not despair. The notion is easily grasped from introductory texts. For actually finding derivatives David Scherfgen throws him or she the perfect lifebuoy in his application program *Derivative Calculator* http://www.derivative-calculator.net/

$$\frac{d}{dx}f(x) = \frac{-\left(cdx^2 - e \cdot bx^2 + 2adx - 2e \cdot x - c + ab\right)}{\left(dx^2 + bx + 1\right)^2}$$

$$\frac{d^2}{d^2x}f(x) = \frac{2\cdot\left(cd^2x^3 - e \cdot bdx^3 + 3ad^2x^2\right.}{\left(dx^2 + bx + 1\right)^3}$$

$$\frac{\left.- 3e \cdot dx^2 - 3cdx + 3abdx - ad - bc + ab^2 + e\right)}{\left(dx^2 + bx + 1\right)^3}$$

f(x) is in ^{O}K units. The first derivative expresses velocity which is the rate of change in function f(x) at any time point x. Second derivatives express rate of change in velocity, called acceleration above the zero line and deceleration below the zero line.

To parameterise the equations we replace the parameters a, b, c, d, e with numerical values taken from the regression records. The numerical value of f(x) for given x yr. TAI is a regression estimate of the Vostok proxy temperature difference in ^{O}K units at x yr. after initiation. Regarding the residual, the symbolic definition is x–f(x). Residuals are measured in ^{O}K units.

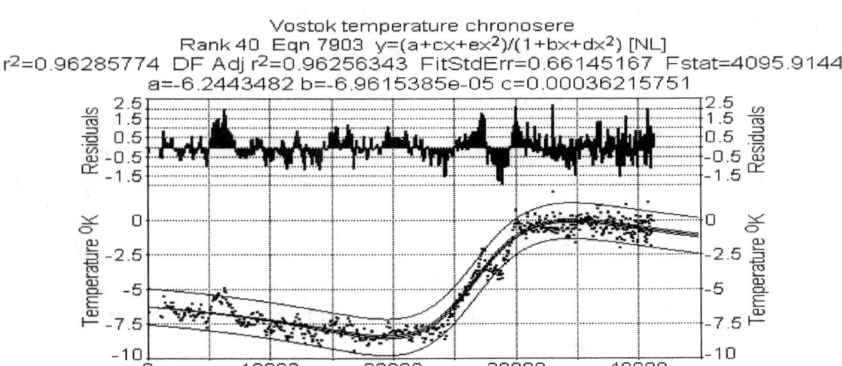

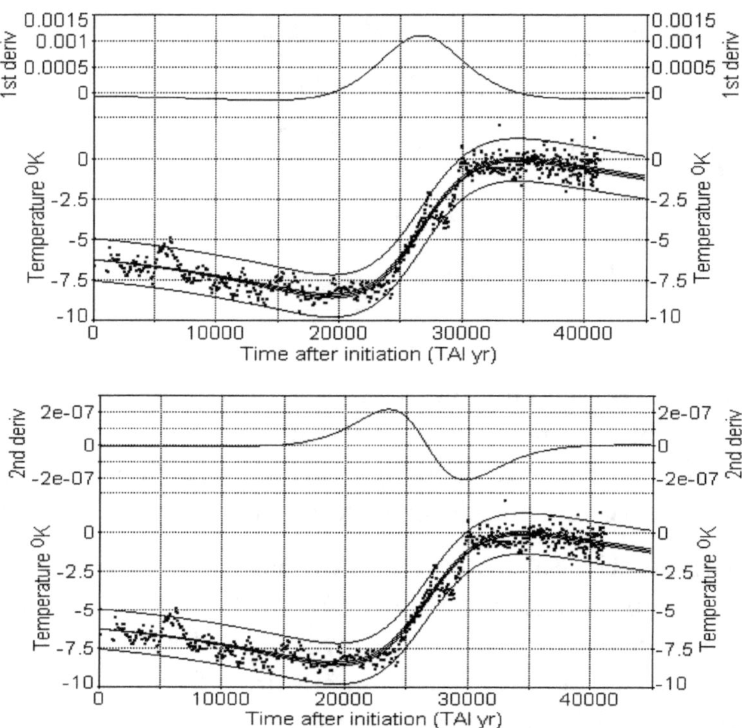

Figure 4B1. Vostok proxy temperature scale. Temperature differences are displayed. The TAI scale indicates elapsed time after initiation (TAI yr.) The initiation is year 41138 on the TBP scale. This is the carbon year when the bottom sediments were laid in the Cwynar sediment core. *Regression line:* centre line imbedded within the 95% confidence belts, one for regression (inner) and the other for prediction (outer). Equation parameters and other regression results are presented in the figure's main caption and in the table below which I copied the tables directly from Table Curve. *Residuals*: deviations of the observed temperature vales from the regression line. *1st derivative:* rate (velocity) of temperature change per year directionally from past to present. *2nd derivative*: acceleration above the zero line, deceleration below the zero line. Numeric table:

Rank 40 Eqn 7903 $y=(a+cx+ex^2)/(1+bx+dx^2)$ [NL]

r^2 Coef Det	DF Adj r^2	Fit Std Err	F-value
0.9628577387	0.9625634260	0.6614516688	4095.9143938

Parm	Value	Std Error	t-value	95% Confidence Limits		P>|t|
a	-6.24434818	2.92606e-10	-2.1341e+10	-6.24434818	-6.24434818	0.0000
b	-6.9615e-05	2.89442e-07	-240.515598	-7.0184e-05	-6.9047e-05	0.0000
c	0.000362158	1.93646e-06	187.0204704	0.000358355	0.000365960	0.0000
d	1.30835e-09	1.15988e-11	112.8007051	1.28557e-09	1.33112e-09	0.0000
e	-5.26e-09	5.68357e-11	-92.5469714	-5.3716e-09	-5.1484e-09	0.0000

Area Xmin-Xmax	Area Precision		
-206866.0585	4.149616e-11		
Function min	X-Value	Function max	X-Value
-8.445643609	19498.682140	-0.068023097	34560.403005
1st Deriv min	X-Value	1st Deriv max	X-Value
-0.000153275	14129.820784	0.0011136232	26745.915088
2nd Deriv min	X-Value	2nd Deriv max	X-Value
-2.10835e-07	29843.920001	2.209376e-07	23613.582716

Singularities [Data Range]
None
Singularities [All Other]
None

Soln Vector	Covar Matrix	SVD Cond
LvMrq/SVD	SVDecomp	2.190565e+21
r^2 Coef Det	DF Adj r^2	Fit Std Err
0.9628577387	0.9625634260	0.6614516688

Source	Sum of Squares	DF	Mean Square	F Statistic	P>F
Regr	7168.1502	4	1792.0375	4095.91	0.0000
Error	276.51157	632	0.43751831		
Total	7444.6617	636			

Description: Vostok temperature chronodsere

X Variable: Time after initiation (TAI yr)

Xmin:	0.0000000000	Xmax:	41138.000000	Xrange:	41138.000000
Xmean:	24384.784929	Xstd:	12084.593776	Xmedian:	26796.000000
X@Ymin:	16775.000000	X@Ymax:	33003.000000	X@Yrange:	16228.000000

Y Variable: Temperature OK

Ymin:	-9.390000000	Ymax:	2.0600000000	Yrange:	11.450000000
Ymean:	-4.000486656	Ystd:	3.4213218208	Ymedian:	-4.020000000
Y@Xmin:	-6.590000000	Y@Xmax:	0.0000000000	Y@Xrange:	6.5900000000

Date	Time	File Source
Mar 15, 2014	9:12:18 AM	CLIPBRD.PRN

The TAI scale has to be explained. Direction problems with the use of the TBP scale in TableCurve forced us to turn to the TAI scale. On the TBP scale progress forward is left. On the TAI scale process forward in the graphs is right.

Renland temperature chronosere
Rank 59 Eqn 7903 $y=(a+cx+ex^2)/(1+bx+dx^2)$ [NL]
r^2=0.88615152 DF Adj r^2=0.88545306 FitStdErr=1.0146305 Fstat=1587.8553
a=-5.2250254 b=-6.4108553e-05 c=0.00030664161

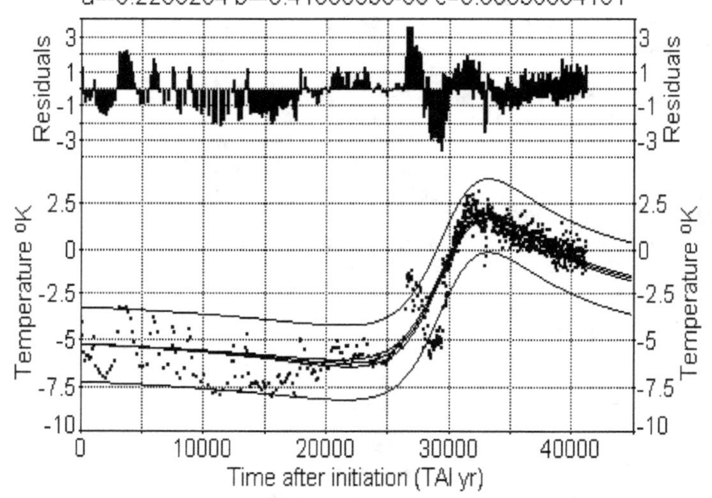

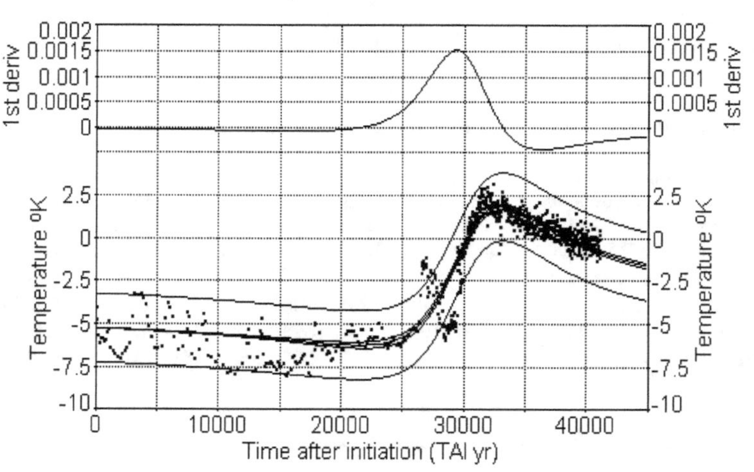

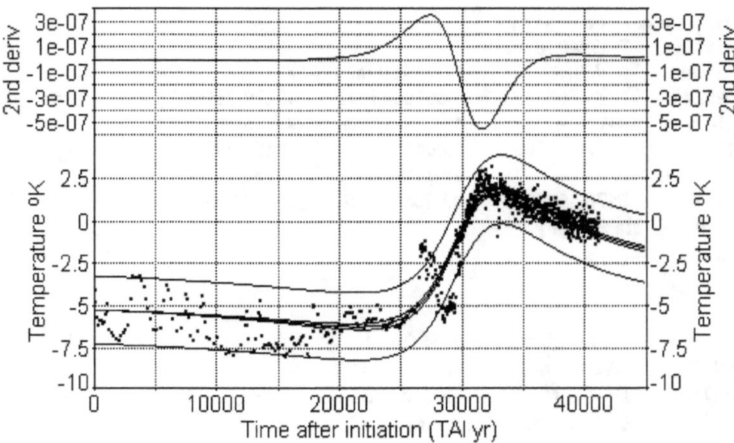

Figure 4B2. Renland proxy temperature differences scale in K unis. The TAI scale is elapsed time after initiation (see remarks following Table 3B1). *Regression line:* centre line imbedded within the 95% confidence belts, one for regression (inner) and the other for prediction (outer). Equation parameters and other regression results are in the figure's main caption and in the table below. *Residuals*: deviations of the observed temperature vales from the regression line. *1st derivative:* rate (velocity) of temperature change along the regression line from past to present per year. *2nd derivative*: acceleration above the zero line, deceleration below the zero line in. Numeric table:

Rank 59 Eqn 7903 $y=(a+cx+ex^2)/(1+bx+dx^2)$ [NL]

r^2 Coef Det	DF Adj r^2	Fit Std Err	F-value
0.8861515183	0.8854530614	1.0146304832	1587.8552532

Parm	Value	Std Error	t-value	95% Confidence Limits		P>|t|
a	-5.22502540	4.13139e-11	-1.2647e+11	-5.22502540	-5.22502540	0.0000
b	-6.4109e-05	2.9233e-07	-219.302040	-6.4682e-05	-6.3535e-05	0.0000
c	0.000306642	7.00685e-07	437.6313130	0.000305266	0.000308017	0.0000
d	1.05607e-09	9.88743e-12	106.8088413	1.03666e-09	1.07547e-09	0.0000
e	-4.4355e-09	2.17592e-11	-203.846208	-4.4782e-09	-4.3928e-09	0.0000

Area Xmin-Xmax	Area Precision		
-154289.6766	7.332481e-16		
Function min	X-Value	Function max	X-Value
-6.214595744	21566.991926	1.8660041142	33270.255617
1st Deriv min	X-Value	1st Deriv max	X-Value
-0.000424688	36379.837551	0.0015239474	29457.035675
2nd Deriv min	X-Value	2nd Deriv max	X-Value
-5.49066e-07	31707.021350	3.543104e-07	27426.750903

Singularities [Data Range]

Singularities [All Other]

None

Soln Vector	Covar Matrix	SVD Cond
LvMrq/SVD	SVDecomp	1.116377e+22
r^2 Coef Det	DF Adj r^2	Fit Std Err
0.8861515183	0.8854530614	1.0146304832

Source	Sum of Squares	DF	Mean Square	F Statistic	P>F
Regr	6538.6293	4	1634.6573	1587.86	0.0000
Error	840.05161	816	1.029475		
Total	7378.6809	820			

Description: Renland temperature chronosere

X Variable: Time sfter initiation (TAI yr)

Xmin:	0.0000000000	Xmax:	41100.000000	Xrange:	41100.000000
Xmean:	30352.395859	Xstd:	9606.5938138	Xmedian:	32900.000000
X@Ymin:	15550.000000	X@Ymax:	32320.000000	X@Yrange:	16770.000000

Y Variable: Temperature OK

Ymin:	-8.000000000	Ymax:	3.1940298510	Yrange:	11.194029851
Ymean:	-1.035177341	Ystd:	2.9997318725	Ymedian:	0.1940298510
Y@Xmin:	-4.000000000	Y@Xmax:	0.0000000000	Y@Xrange:	4.0000000000

Date	Time	File Source
Mar 2, 2014	4:33:17 PM	CLIPBRD.PRN

Comparison of chronoseres

The question is this: shall we regard the Vostok and Renland temperature chronoseres as variants of the same global model $f(x)=\frac{a+cx+ex^2}{1+bx+dx^2}$ or consider them different? I believe Vostok and Renland are variants of the same model for shape but not for scale.

Taking the results we already have, we can say much regarding the properties of $f(x)=\frac{a+cx+ex^2}{1+bx+dx^2}$:

1. The fit to the data is almost perfect in both cases, R^2=0.963 for Vostok and R^2=0.886 for Renland.

2. We can read from the numeric tables below (copied directly from the previous sections) that each of the parameters a, b, c, d, and e is statistically significant (see P>|t| approaching zero). Furthermore, the parameters are significantly different between the Vostok and Renland chronoseres (the 95% confidence limits do not overlap):

Vostok statistic:

Parm	Value	Std Error	t-value	95% Confidence Limits		P>|t|
a	-6.24434818	2.92606e-10	-2.1341e+10	-6.24434818	-6.24434818	0.00000
b	-6.9615e-05	2.89442e-07	-240.515598	-7.0184e-05	-6.9047e-05	0.00000
c	0.000362158	1.93646e-06	187.0204704	0.000358355	0.000365960	0.00000
d	1.30835e-09	1.15988e-11	112.8007051	1.28557e-09	1.33112e-09	0.00000
e	-5.26e-09	5.68357e-11	-92.5469714	-5.3716e-09	-5.1484e-09	0.00000

Renland statistics R1

Parm	Value	Std Error	t-value	95% Confidence Limits		P>|t|
a	-5.22502540	4.13139e-11	-1.2647e+11	-5.22502540	-5.22502540	0.0000
b	-6.4109e-05	2.9233e-07	-219.302040	-6.4682e-05	-6.3535e-05	0.0000
c	0.000306642	7.00685e-07	437.6313130	0.000305266	0.000308017	0.0000
d	1.05607e-09	9.88743e-12	106.8088413	1.03666e-09	1.07547e-09	0.0000
e	-4.4355e-09	2.17592e-11	-203.846208	-4.4782e-09	-4.3928e-09	0.0000

Renland statistics R2 (more regression results on this are not given in this essay)

Parm	Value	Std Err	t-value	UL	LL	P>|t|
a	-15.91258	1.0534E-09	-15106000000	-15.91258	-15.91258	0.0000
b	-6.41E-05	2.9233E-07	-219.302039	-6.47E-05	-6.35E-05	0.0000
c	0.0009339	2.1339E-06	437.6313647	0.0009297	0.0009381	0.0000
d	1.056E-09	9.8874E-12	106.808841	1.037E-09	1.075E-09	0.0000
e	-1.35E-08	6.6267E-11	-203.846219	-1.36E-08	-1.34E-08	0.0000

3. To illuminate the models shape/scale properties we need the parameter values and the values in columns under R1/V, Renland R2, R2/V in the following:

Parameter	Vostok V	Renland R1	R1/V	Renland R2	R2/V
α	6.06	0.67	--	0.22	--
a	-6.24435	-5.22503	0.836761	-15.9126	2.548317
b	-6.96E-05	-6.41E-05	0.920908	-6.41E-05	0.920908
c	0.000362	0.000307	0.846708	0.000934	2.578607
d	1.31E-09	1.06E-09	0.807177	1.06E-09	0.807177
e	-5.26E-09	-4.44E-09	0.843251	-1.35E-08	2.568061

The α values are arbitrary scaling coefficient applied to the raw temperature data or equivalently to a, c and e. In Vostok α=6.06, in Renland 1 α=0.67, and under Renland 2 α=0.22. The values in the V, R1 and R2 columns come directly from regression analysis. R/V columns reveal something very interesting. Under R1/V the fractions are very close for each parameter. But when we change scale from α=0.67 to α=0.22 two parameter groups are isolated. One contains a, c, e, the parameters sensitive to scale change, and the other contains b and d, the parameters that are scale independent. Parameters remaining unaffected by scale change are the parameters of shape. We conclude that function $f(x)=\frac{a+cx+ex^2}{1+bx+dx^2}$ successfully isolates the scale effect by the numerator $a+cx+ex^2$ from the shape effect $1+bx+dx^2$ by the denominator. Since the shape parameters b, d do not isolate Vostok from Renland we conclude that they are scaled variants of the same global model $f(x)=\frac{a+cx+ex^2}{1+bx+dx^2}$.

The scale and shape functions for Vostok and Renland generated the graphs in the following figures:

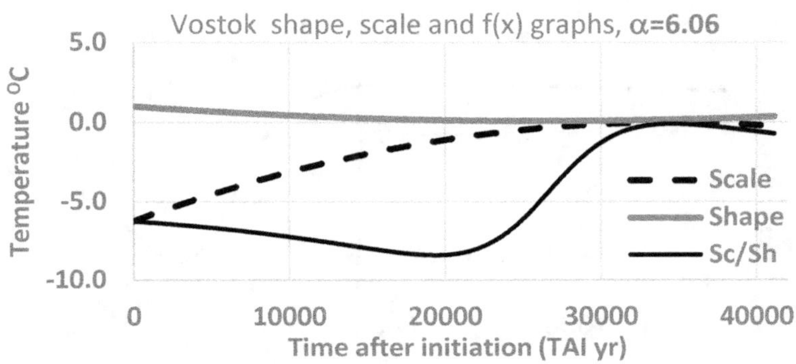

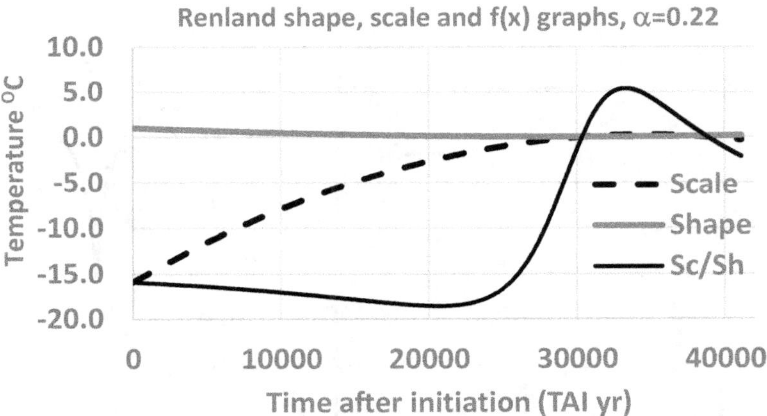

The shape function has curve on top in each graph, scale function middle, and the map of f(x) = Sc/Sh bottom. It is quite clear that the shape effect is scale independent, i.e. not affected by the diminishing temperature difference as time approaches the present. The scale effect is a direct function of the temperature difference.

The following two graphs are generated by percentages,

$$Sc_{\%} = 100\frac{Sc}{Sc+Sh} \quad \text{and} \quad Sh_{\%} = 100\frac{Sh}{Sc+Sh}$$

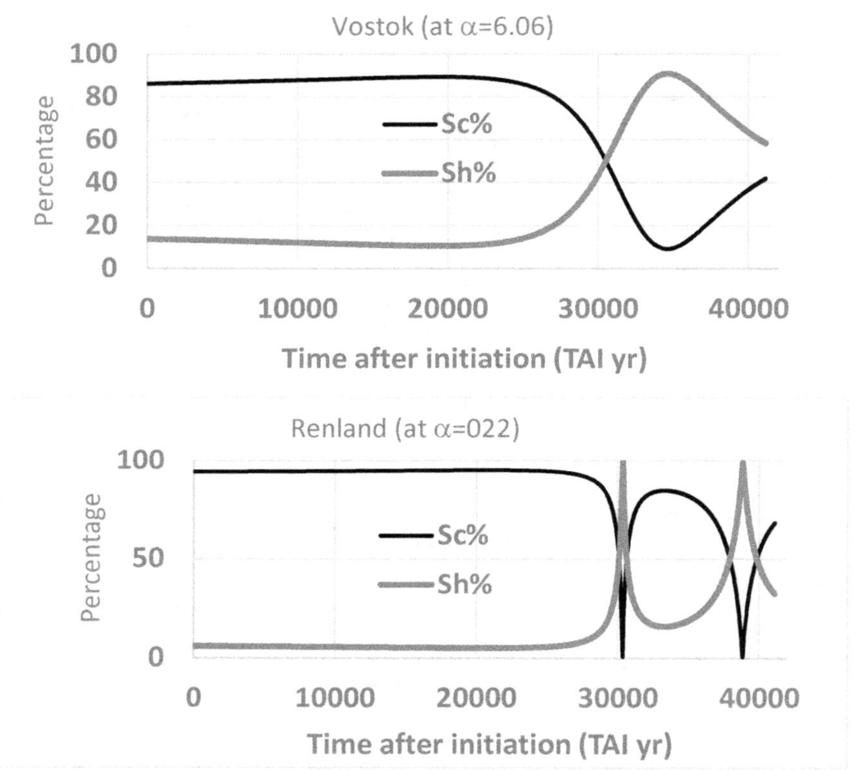

What can we say about graphs Sc% and Sh%? These graphs are very similar up to about year 30340 TAI or 11000 TBP, the point at which Sc and Sh change dominance. In the Vostok chronosere this means more or less the end of the Sc/Sh story. In the Renland chronosere the change of dominance is temporary and then repeated at year 38860 TAI. Discounting the last 11 k yr. portion on the time scale, the Vostok and Renland chronoseres perform similarly.

4. Extreme and central values - maxima, minima, inflexion points, mean - clarify further the suitability of f(x) to isolate the shape and size effects in highly practical terms:

Extreme points of $f(x)=\frac{a+cx+ex^2}{1+bx+dx^2}$	Vostok kyr. TAI or ^{O}K	Renland kyr. TAI or ^{O}K	|V-R| kyr. or ^{O}K	|V/R|
x of f(x) min yr.	19.5	21.6	2.1	0.9
minimum ^{O}K	-8.446	-18.926	10.480	0.45
x of f(x) max yr.	34.6	33.3	1.3	1.04
maximum ^{O}K	-0.068	5.683	5.751	0.012
x of 1st deriv min yr.	14.1 or 37.0	36.4	0.6	1.02
x of 1st deriv max yr.	26.7	29.4	2.7	0.91
x of 2nd deriv min yr.	29.8	31.7	1.9	0.94
x of 2nd deriv max yr.	23.6	27.4	3.8	0.86

It is not difficult to see that the Vostok and Renland chronoseres differ regarding the size indicators (^{O}K, lines 2, 4, 6), but they are similar in terms of the x coordinates of the extreme values. These are telling us that we should accept the single global model conjecture for shape.

Having reached this point in testing, I feel liberated to couple the Cwynar plant particle chronosere with the Vostok temperature chronosere in quantum analysis, and leave the Renland data for considerations to others.
Other reasons for choosing Vostok becomes obvious when I point out what amounts to a major imbalance in the sampling design at the Renland research site. The two graphs below tell the story:

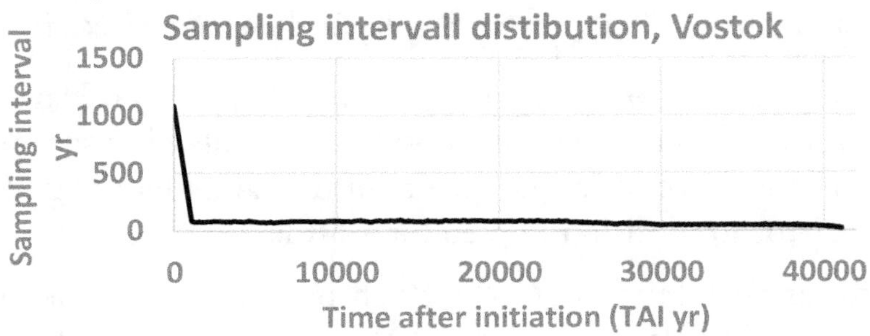

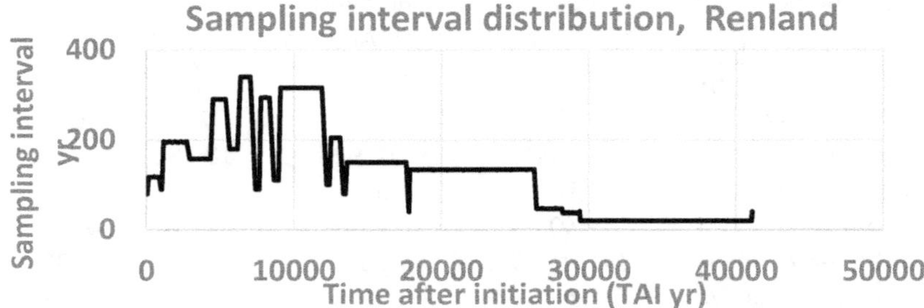

The Vostok sampling interval size is kept similar for the entire length of the chronosere, except for a blip at initiation. The Renland site has a very unbalance sampling interval distribution. What makes the design even less appealing to me is its low sapling intensity in the lower 25 kyr. portion of the ice core (see original data).

At this point I return to the questions I posed earlier:

1. Do I plan to use both temperature chronoseres? I opted for Vostok. I justify my choice on grounds of relevance and also for being a statistically a better sample.

2. The two chronoseres obviously differ, but the differences are more related to scale than shape. How to isolate the two aspects and which should weigh more heavily when making a choice? The parameters in $f(x)=\frac{a+cx+ex^2}{1+bx+dx^2}$ isolated the shape effect from the scale effect. I give more weight for the shape effect. Having found little difference between Vostok and Renland in that regard and I felt free to let the quality of the sampling design be my mediator for decision I opted for Vostok.

3. Is the arctic location of Cwynar's chronosere not mandating the use of the Renland chronosere? No, it does not. The chronoseres were shown to be variants of the same global temperature

model for shape and my main interest is in shape. I offer further comments. Some expand on what I already said:

4, I emphasised in an earlier paper (Orlóci 2008)[23] that the Vostok temperature chronosere is globally generalised and can be projected to global sites by proxy. The foregoing sections supplied new justification for this. How to project the Vostok temperature chronosere to other global sites? Schweingruber's (1996) graph can help. It gives the latitudinal gradient for deuterium on temperature as a perfectly straight line. This facilitates the conversion of Vostok ground level temperature values to globally generalised values (Orlóci 2008). But this transformation will add to and not diminish complications inherent in deriving the temperature values by proxy under a rather uncontrolled scale effect, albeit the scale is not haphazard.

Further on capturing shape

Before we move on to the next topic we should recall that in the analysis so far our main concern was the isolation of directed trend from residuals. "Directed" implies anything that is not "random". The function to capture the directed trend was our choice. We needed to inspect critical properties of many functions before we found one of fit and right shape, not unlike what we thought Nature could produce.

We relied on R^2 (the coefficient of determination) for ranking the functions for closeness of fit. R^2 is a complementary function of the residuals. As the fit gets better, the residuals get smaller and R^2 increase. But this does not mean that the crispness of the trend increases also from an ecologist's point of view.

[23] László Orlóci 2008 -- http://www.akademiai.com/content/l246056112n4/?k=Orloci

R^2 is simply not sufficient to make a good choice. It does not indicate to us how much natural is the function's shape. Consider the portrayal of Cwynar's plant particle chronosere by different function:

1. Fourier series polynomial

Che/Am chronosere

Rank 1 Eqn 6850 Fourier Series Polynomial 10x2

r²=0.89096383 DF Adj r²=0.87033537 FitStdErr=3.0310378 Fstat=45.759106

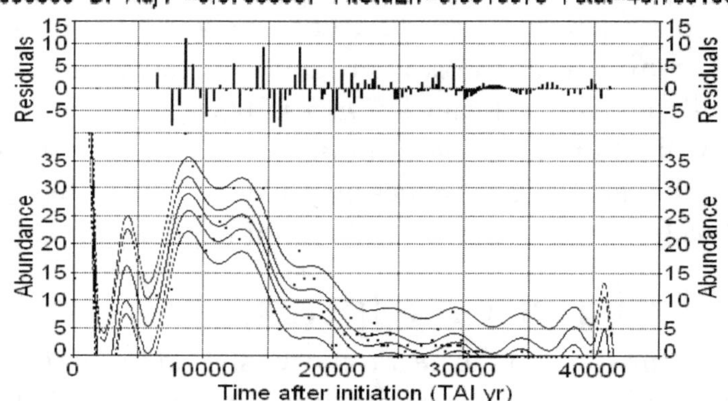

2. Cauchy-Lorentz equation

Che/Am chronosere

Rank 77 Eqn 8004 [Lorentzian] y=a+b/(1+((x-c)/d)2)

r²=0.83673115 DF Adj r²=0.831629 FitStdErr=3.4559853 Fstat=220.36928

Residuals
Abundance
Time after initiation (TAI yr)

The captions include vital information. Note how meticulously the Fourier series polynomial picks out the finest details. But I cannot imagine Nature being a comparably meticulous technical artist. I hardly can see cases where I would be interested in such fine details when I look for generalizable trends. The smoother trend line of the Cauchy-Lorentz function and its overall shape appeal to me to be much more natural. I summarise my resolution of the best fit versus natural shape paradox:

1. Different functions may have the same or similar values of R^2. From among these the simplest is of specific interest to me.

2. The function I chose in the example implied the some loss of precision regarding the mechanistic criteria of fit. I am prepared to give up even more precision if needed to satisfy *my* criteria of "natural shape". Such a resolution of the paradox has lead me to two functions. I used one already for the Vostok and Renland chronoseres: $f(x)=\frac{a+cx+ex^2}{1+bx+dx^2}$. The other is the Cauchy-Lorentz function just introduced: $f(x)=\frac{b}{\frac{(x-c)^2}{d^2}+1}+a$

More on extreme points

The functions f(x) in both cases are differentiable. I have special interest in the 1st and 2nd derivatives. The derivatives identify points where trends change and mark out segments in the chronosere which can have unique ecological significance. These are similar traits comparable between chronoseres.

Figure 4A, B1 and B2 revisited

The functions f(x) and its derivatives free us from having to deal with the raw graphs (Figure 4A). The graphs in Figure 4B support further discussion of what I consider the key traits of the Vostok temperature chronosere:

1. *Shape.* The mirror image of the fitted f(x) on the TAI scale is like a Z placed tightly within the 95% confidence belt for regression which in turn is packed somewhat less tightly within the 95% confidence belt for prediction.

2. *Residuals.* We see a regularity what appear random oscillation and chance clumping by sign in the graphs. I did not test the randomness of clumping. Nor did I checked if this is some ways connected to the chronologically uneven time steps in the sampling. If the clumping pattern is random, we can say that the trend extracted is crisp.[24]

3. *Natural chronosere segments.*

There are four segments isolated by the 1st derivatives extreme points:

a. 0 yr. to19kyr. AI; climate cooling is bottoming out.

b. 19 kyr. to 27k yr. AI; warming cycle underway at an accelerated rate.

c. 27kyr. to 35k yr. AI; major inflexion point reached in the warming process, the rate of temperature increase declining.

d. 35k AI to present; climate warming passes maximum and a slow cooling trend takes effect.

The chronoseres segmentation based on the 2nd derivative shift the segments somewhat back in time.

---

[24] One more comment is in order. What we lump under the aegis of "residuals" even in a truly random state, manifests much more than pure natural stochasticity. The "residuals" category is in fact bloated by measurement errors, which can be determined if certain conditions existed at time of measurement, and sampling errors, which can be estimated if certain formal conditions existed in the course of sampling. A question to the reader with experience in the applications of regression analysis: how often did this reader isolate the three components of random residuals?

4. *Precision of fit.* The R^2 and standard error give us direction in this. Complete list of numerical values is in the numeric tables following Figures 4B1 and 4B2.

6. *Isolation of trend and random oscillations.* A main objective of regression analysis is completed with the extraction of the residuals. The success of the exercise really depends not just on how large is R^2 and how small is the value of the standard error, in other words the level of precision. What matters very much is the probability that such a level of precision can arise by pure chance in sampling of a totally random universe. To judge this the t and F values have to be scrutinised. When they indicate that we have a very low probability event under the random universe assumption, we can declare success.

7. *Isolation of size and shape components of f(x).* We found utility of isolation in the discovery of a global regularity: the Vostok chronosere and the Renland chronosere are scale variants of the same global model.

There are many ways that a regression analysis calls upon personal judgment in making choices, setting probability limits or just simply deciding if what is done is the right thing to do. Much of the biasedness comes because we have no theory which mandate a specific f(x) a priori. It is a paradox that exactly the lack of a biding theory makes the exercise most fascinating. In fact that lack offers the opportunity to probe for and discover intrinsic laws a posteriori, such as the functional form of the process in the unfolding of the Vostok chronosere through more than 40 millennia.

What I discussed in this chapter sets the *modus operandi* in a large number of regression cases that the reader will encounter in the main text.

Beringian vegetation

Hultén (1937) emphasises the role of Beringia's vegetation mosaic as a refugium where the northern flora survives in time of glacial advances and from which species migrate out to repopulate vast areas of Terra Nova in sites freed up after glacial retreat.

Beringia's arid, arctic climate supported the broad, zonal distribution of the cold steppe during the climate cooling cycle. The leading plant families in the cold steppe include:

Poaceae (Festuca, Luzula, Artagrostis)
Asteracea (Artemisia)
Chenopodiaceae
Amaranthaceae
Fabaceae (Oxytropis
Rosaceae (Potentilla)

The steppe is a component of the current mosaic pattern which is an admixture of tundra, taiga, and boreal forest types.

The performance graphs of selected palynomorph taxa in Figure 5A1 are based on the Cwynar records. The sequence in which the graphs are presented signal climate change in the Hanging Lake site from the early cold steppe epoch (right) to the present state of the current mosaic of vegetation types (left). The Vostok temperature differences graph is included for quick reference.

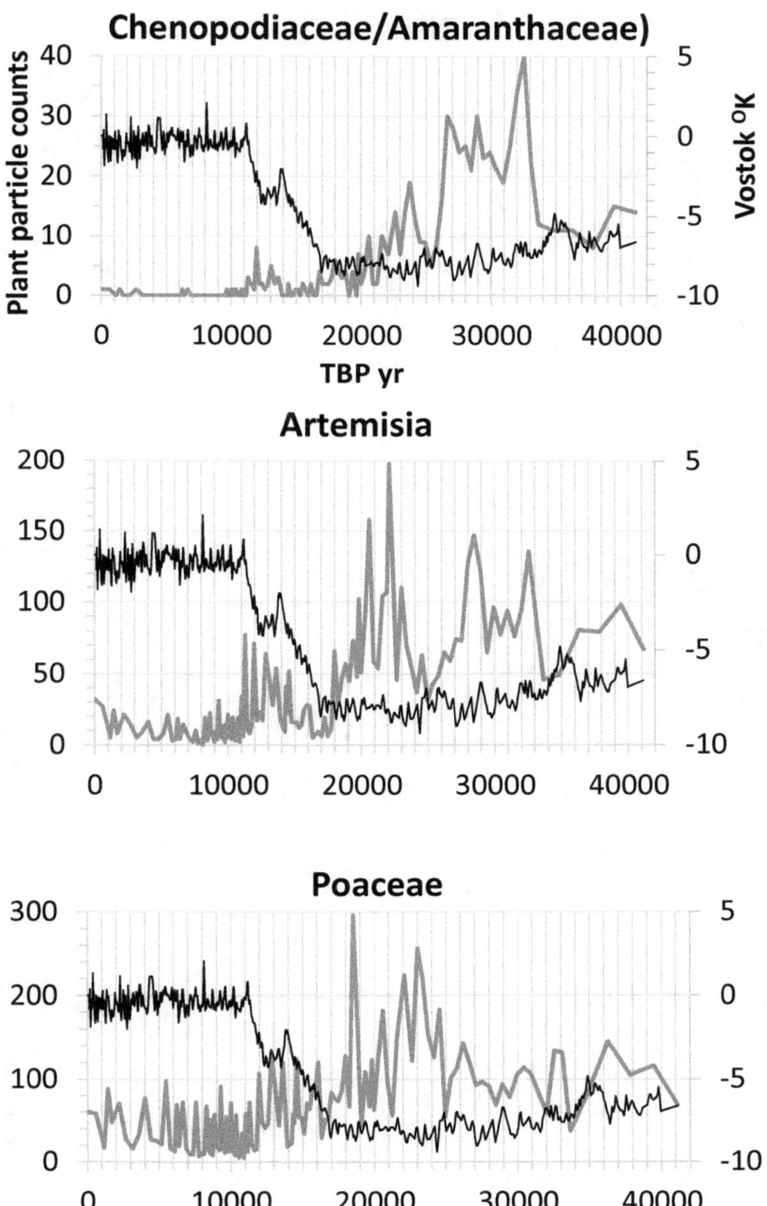
Chenopodiaceae/Amaranthaceae)
Plant particle counts
Vostok °K
TBP yr
0
10000
20000
30000
40000
40
30
20
10
0
5
0
-5
-10
Artemisia
200
150
100
50
0
5
0
-5
-10
0
10000
20000
30000
40000
Poaceae
300
200
100
0
5
0
-5
-10
0
10000
20000
30000
40000

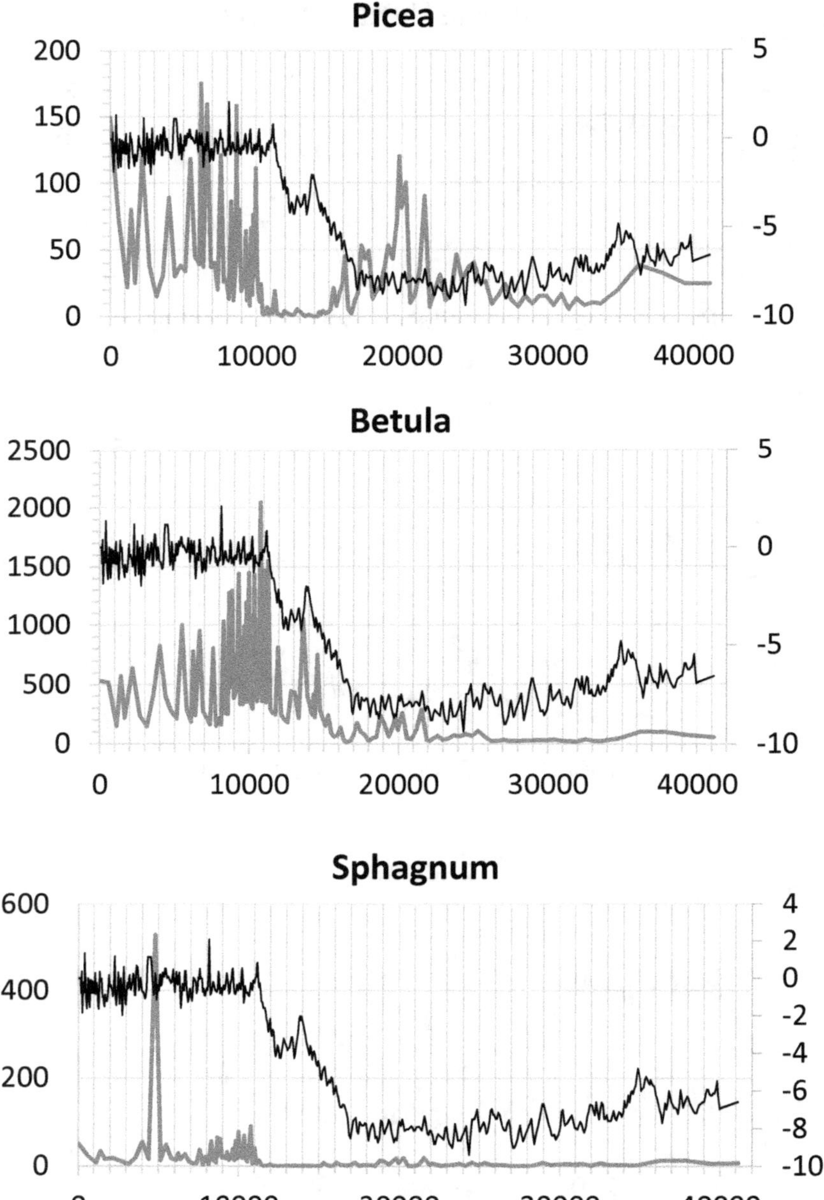
Picea
200
150
100
50
0
5
0
-5
-10
0
10000
20000
30000
40000
Betula
2500
2000
1500
1000
500
0
5
0
-5
-10
0
10000
20000
30000
40000
Sphagnum
600
400
200
0
4
2
0
-2
-4
-6
-8
-10
0
10000
20000
30000
40000

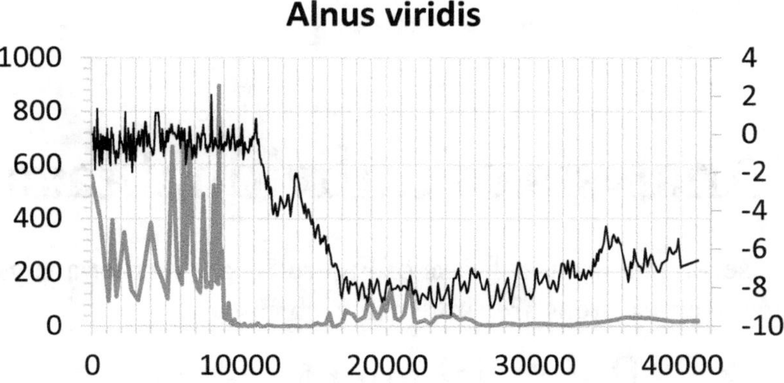

Figure 5A. Lower graph in each figure represents a 133-valued chronosere of the selected palynomorph taxon. The Vostok temperature graph (637 points) is included for quick reference. Observe the change in the Hanging Lake site from the arid cold steppe climate on right to the present much warmer climate on left. Attention is drawn to the palynomorph taxa of unequal status in a species based taxonomy.

Analysis of individual taxa

The seven palynomorph taxa of Figure 5A are further analysed. The analytical steps are similar in each case.

Potential energy levels

Max Planck's energy based entropy equation $nH = \ln C$ is applied. C is parameterised by the total particle count (energy units counts) T and the number of paleorelevés n in each taxon. We have:

Taxa	Che/Am	Artem	Poac	Picea	Betula	Sphag	Alnus
n	79	132	133	129	133	121	123
T	691	4970	9370	4278	44137	2234	11581
nH	254.68	612.68	699.84	582.61	905.23	477.03	682.68
H	3.2238	4.6415	5.262	4.5164	6.8062	3.9424	5.5502
P_a	0.0398	0.0096	0.0052	0.0109	0.0011	0.0194	0.0039
P_b	0.9602	0.9904	0.9948	0.9891	0.9989	0.9806	0.9961
w_a	0.0016	0.0001	0	0.0001	0	0.0004	0
w_b	0.922	0.9808	0.9897	0.9783	0.9978	0.9616	0.9922
w_{ab}	0.0764	0.0191	0.0103	0.0216	0.0022	0.0381	0.0077
w_{ab}%	15.2878	3.82	2.0633	4.3237	0.4423	7.6103	1.5486

The values in the last two rows of the table indicate the instability levels. Conversion to stability levels is in the manner of $0.5\text{-}w_{ab}$ and 200(0.5-wab)%. The potential energy based stability levels are very high.

We now turn to regression analysis of the oscillograms. I give more details in the Che/Am case and less in the other cases. The following include suggestions in the interest of avoiding ambiguities when viewing the graphs:

1. *Chronosere direction.* Read carefully the labels. Identify the axes. Remember, if the time scale is TBP the zero of the scale indicates "present" and the natural direction of the chronosere is from right to left. If the time scale is TAI the zero point is initiation and the natural direction of progress is lined up from left to right.

2. *Concepts, abbreviations.* Concepts have to be clear. Parallel study of the references already given and others may be needed. The abbreviations and terms used in the numeric tables are explained under "Numeric summary" in the alphabetically arranged index of the Help Topics on SYSTAT'.s web page:

http://systat.co.kr/products/TableCurve2D/resources/help.html

3. *Extreme points.* The taxon's estimated optimum point on the time axes is at the functions maximum. Follow the optima drifting as it were from past to present for the successive taxa. Identify all extreme points and use them to set intervals. Look for coincidences of spatial patters delineated by the extreme points with those on the Vostok temperature chronosere.

6. *Statistical parameters.* These help to evaluate the closeness of fit and uniqueness of the models.

Chenopodiaceae/Amaranthaceae

The species of the families Chenopodiaceae and Amarantaceae are characteristic components in the Beringian arid cold steppe. By extension we consider the Che/Am palynomorph taxa characteristic of the cold steppe. The function fitted to the Che/Am oscillogram of Figure 5A is the Cauchy-Lorentz equation,

$$f(a,b,c,d,x) = \frac{b}{\frac{(x-c)^2}{d^2}+1}+a$$

This has 1st and 2nd derivatives

$$\frac{d}{dx}f(a,b,c,d,x) = \frac{-2bd^2\cdot(x-c)}{\left((x-c)^2+d^2\right)^2}$$

$$\frac{d^2}{d^2x}f(a,b,c,d,x) = \frac{2bd^2\cdot\left(3x^2-6cx-d^2+3c^2\right)}{\left(x^2-2cx+d^2+c^2\right)^3}$$

The graphs are drawn in Figure 5B. The figure also include a graph of residuals in the manner of $f(x)_{observed} - f(x)_{fitted}$.

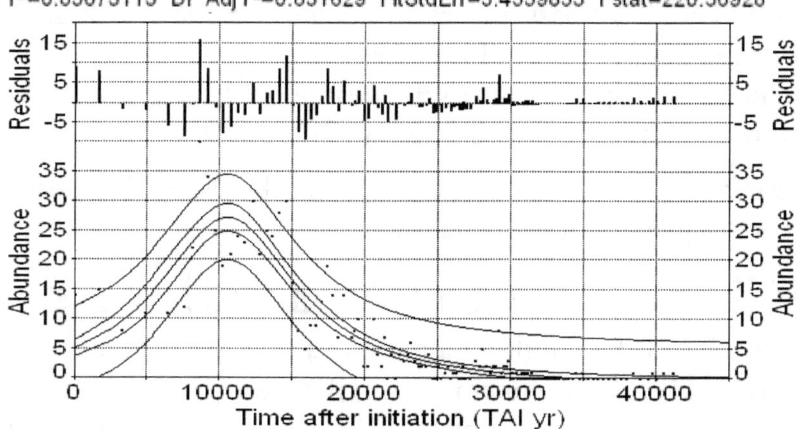

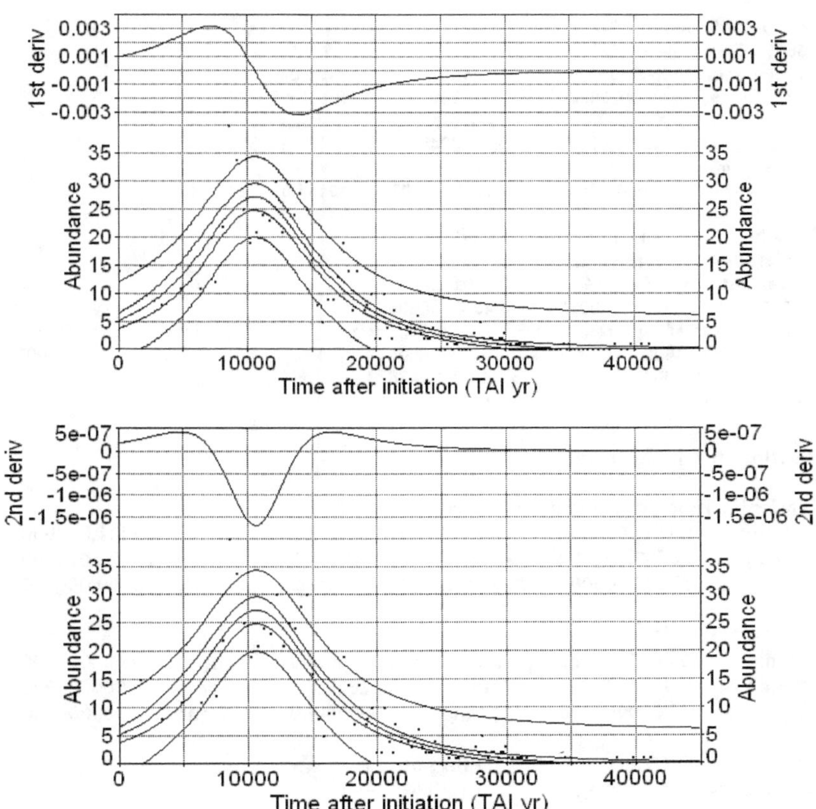

Figure 5B. The Cauchy-Lorentz trend line fitted to the Che/Am oscillogram of Figure 5A is in the centre within two inclusive 95% confidence belts, one for regression and the other for prediction. Graphs of residuals, 1st derivative and 2nd derivative are identified. The numeric table:

Rank 77 Eqn 8004 [Lorentzian] $y=a+b/(1+((x-c)/d)^2)$

r^2 Coef Det	DF Adj r^2	Fit Std Err	F-value
0.8367311715	0.8316290206	3.4559850698	220.36931784

Parm	Value	Std Error	t-value	95% Confidence Limits		P>|t|
a	-1.55475029	0.591905774	-2.62668545	-2.72585036	-0.38365022	0.00967
b	28.75475836	1.178683326	24.39566058	26.42270453	31.08681219	0.00000
c	10606.98709	284.2726271	37.31272757	10044.54673	11169.42744	0.00000
d	5827.391239	517.7299848	11.25565722	4803.049784	6851.732694	0.00000

Area Xmin-Xmax	Area Precision		
346679.42772	3.983727e-09		
Function min	X-Value	Function max	X-Value
-0.543955749	41137.000000	27.200008071	10606.991418
1st Deriv min	X-Value	1st Deriv max	X-Value
-0.003204996	13971.433731	0.0032049956	7242.5418654
2nd Deriv min	X-Value	2nd Deriv max	X-Value
-1.69352e-06	10606.965011	4.233655e-07	16454.823221

Procedure	Minimization	Iterations
LevMarqdt	Least Squares	9
r^2 Coef Det	DF Adj r^2	Fit Std Err
0.8367311715	0.8316290206	3.4559850698

Source	Sum of Squares	DF	Mean Square	F Statistic	P>F
Regr	7896.1629	3	2632.0543	220.369	0.00000
Error	1540.7544	129	11.943833		
Total	9436.9173	132			

Description: Ch/Am cronosere

X Variable: Time after initiation (TAI yr)

Xmin:	0.0000000000	Xmax:	41137.000000	Xrange:	41137.000000
Xmean:	25563.601504	Xstd:	9209.0332883	Xmedian:	27522.000000
X@Ymin:	27222.000000	X@Ymax:	8602.0000000	X@Yrange:	18620.000000

Y Variable: bundance

Ymin:	0.0000000000	Ymax:	40.000000000	Yrange:	40.000000000
Ymean:	5.1954887218	Ystd:	8.4552822351	Ymedian:	1.0000000000
Y@Xmin:	14.000000000	Y@Xmax:	1.0000000000	Y@Xrange:	13.000000000

Date	Time	File Source
Feb 14, 2014	5:53:24 PM	CLIPBRD.PRN

It is quite obvious from the regression results that the Che/Am oscillogram in Figure 5A has trend and residuals tightly convoluted. These are isolated in Figure 5B. An overview of extreme and central values, such as maxima, minima, inflexion points, and mean, should clarify further the suitability of f(x) to isolate the shape and sise effects in practical terms:

Extreme points of $f(x)=\dfrac{b}{(x-c)^2/d^2+1}+a$	Vostok kyr. TAI or abundance
x of f(x) min yr.	41.1

minimum abund	0
x of f(x) max yr.	10.6
maximum abund	27.2
x of 1st deriv min yr.	14.0
x of 1st deriv max yr.	7.2
x of 2nd deriv min yr.	10.6
x of 2nd deriv max yr.	16.5

What does this table tell us?

1. The Cauchy-Lorentz function puts maximum abundance at x=10607 yr. on the TAI scale (or 30531 yr. on the TBP scale) well within the grate climate cooling cycle into the epoch the arid cold step.

2. The value of the 1st derivative falls to minimum at x=13971 yr. AI (or 27167 yr. BP), and attains maximum at x=7242 yr. AI (or 33896 yr. BP). These points on the time axis bracket the optimal temperature range of Che/Am.

3. The 2nd derivative has three significant points: minimum at x=10607 yr. AI, where the Cauchy-Lorentz function reaches maximum; maximum at x=16455 yr. AI; and secondary maximum around 5000 yr. AI. The maxima bracket the optimum range as well.

Ch/Am plant particle chronosere
Rank 1 Eqn 8160 [Line Robust None, Gaussian Errors] y=a+bx
r^2=0.36235337 DF Adj r^2=0.35254342 FitStdErr=6.777499 Fstat=74.442943
a=-0.48595561
b=-1.4554898

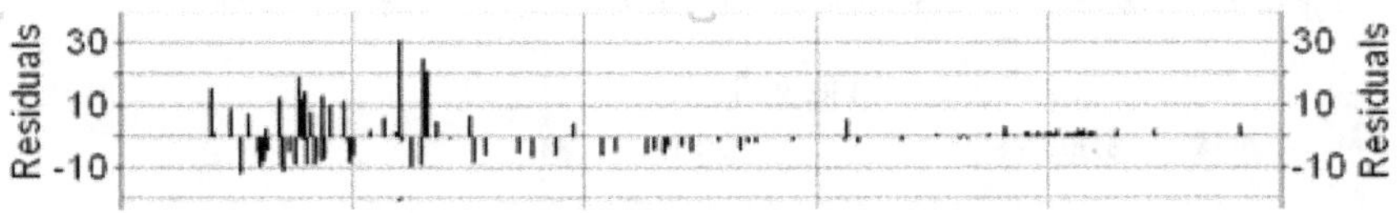

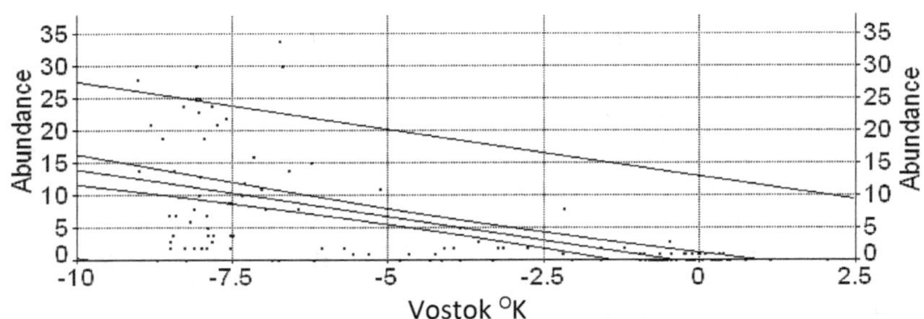

Figure 5C. Linear regression of the Che/Am abundance chronosere with respect to Vostok temperature differences. Numeric table:

Rank 1 Eqn 8160 [Line Robust None, Gaussian Errors] y=a+bx

r^2 Coef Det	DF Adj r^2	Fit Std Err	r^2 Attainable
0.3623533706	0.3525434225	6.7774989778	0.9614810654

Source	Sum of Squares	DF	Mean Square	F Statistic	P>F
Regr	3419.4988	1	3419.4988	74.4429	0.00000
Error	6017.4185	131	45.934492		
Total	9436.9173	132			
Lack Fit	5653.9185	119	47.51192	1.56848	0.19353
Pure Err	363.5	12	30.291667		

Description: CLIPBRD.PRNChe/Am chronosere

X Variable: Vostok oK

Xmin:	-9.020000000	Xmax:	2.0600000000	Xrange:	11.080000000
Xmean:	-3.903458647	Xstd:	3.4969151005	Xmedian:	-3.850000000
X@Ymin:	-2.060000000	X@Ymax:	-6.990000000	X@Yrange:	4.9300000000

Y Variable: Abundance

Ymin:	0.0000000000	Ymax:	40.000000000	Yrange:	40.000000000
Ymean:	5.1954887218	Ystd:	8.4552822351	Ymedian:	1.0000000000
Y@Xmin:	28.000000000	Y@Xmax:	0.0000000000	Y@Xrange:	28.000000000

Date	Time	File Source
Feb 14, 2014	6:00:28 PM	CLIPBRD.PRN

In Figure 5C the Che/Am chronosere is projected as a linear function of the Vostok temperature chronosere. We can see the high deviation values displaced far into the climate cooling cycle. This suggests maximum dispersion of Che/Am under the cold steppe climate.

Artemisia

Artemisia plant particle chronosere

Rank 166 Eqn 8004 [Lorentzian] $y=a+b/(1+((x-c)/d)^2)$

$r^2=0.55763504$ DF Adj $r^2=0.54381113$ FitStdErr=25.854486 Fstat=54.204806

a=-21.30527 b=114.75181

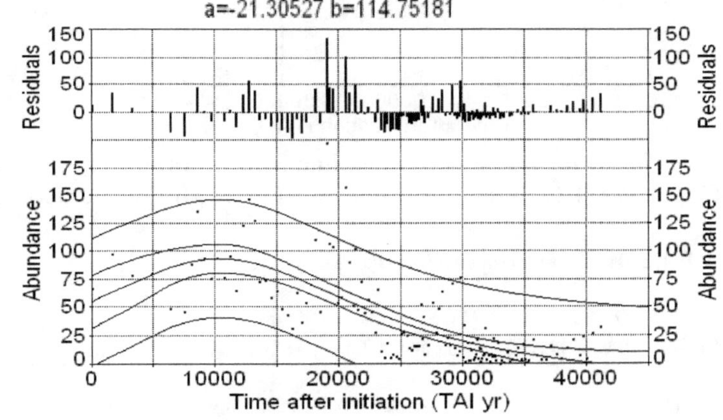

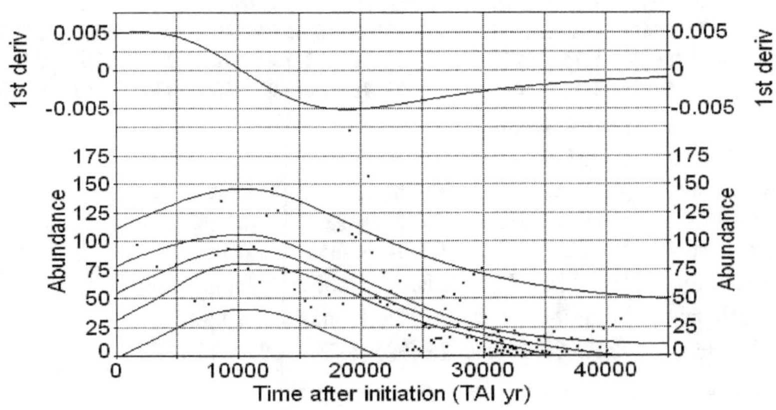

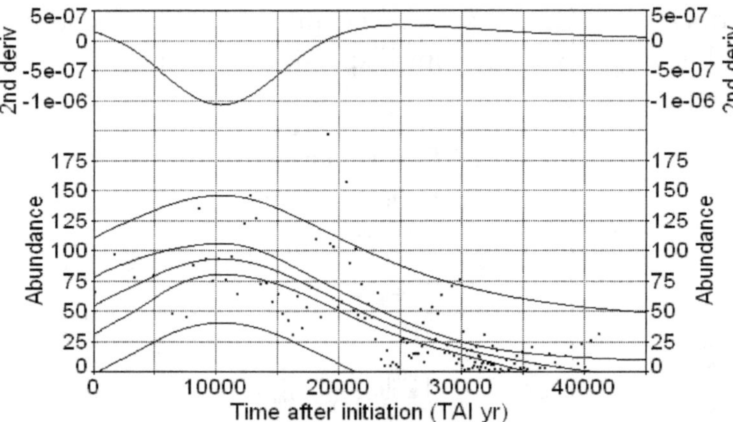

Figure 5D1 The Cauchy-Lorentz trend line fitted to the Artemisia oscillogram of Figure 5A. Graph types are same as in Figure 5B. The numeric table:

Rank 166 Eqn 8004 [Lorentzian] $y=a+b/(1+((x-c)/d)^2)$

r^2 Coef Det	DF Adj r^2	Fit Std Err	F-value
0.5576350390	0.5438111340	25.854485604	54.204805512

Parm	Value	Std Error	t-value	95% Confidence Limits		P>|t|
a	-21.3052700	14.39669625	-1.47987216	-49.7894862	7.178946098	0.14135
b	114.7518147	13.28453401	8.638000746	88.46803878	141.0355906	0.00000
c	10431.18824	1275.570259	8.177666556	7907.441246	12954.93524	0.00000
d	14688.75206	3341.775478	4.395493402	8076.967433	21300.53669	0.00002

Area Xmin-Xmax	Area Precision		
2059990.8918	2.44215e-10		
Function min	X-Value	Function max	X-Value
0.0630535963	41138.000000	93.446544628	10431.187763
1st Deriv min	X-Value	1st Deriv max	X-Value
-0.005074188	18911.752199	0.0050741880	1950.6332864
2nd Deriv min	X-Value	2nd Deriv max	X-Value
-1.0637e-06	10430.987665	2.659254e-07	25120.004675

Procedure	Minimization	Iterations
LevMarqdt	Least Squares	13
r^2 Coef Det	DF Adj r^2	Fit Std Err
0.5576350390	0.5438111340	25.854485604

Total 194930.95 132

Description: Artemisia plant particle chronosere

X Variable: Time after initiation (TAI yr)

Xmin:	0.0000000000	Xmax:	41138.000000	Xrange:	41138.000000
Xmean:	25563.609023	Xstd:	9209.0461001	Xmedian:	27522.000000
X@Ymin:	33006.000000	X@Ymax:	19043.000000	X@Yrange:	13963.000000

Y Variable: Abundance

Ymin:	0.0000000000	Ymax:	198.00000000	Yrange:	198.00000000
Ymean:	37.368421053	Ystd:	38.428499857	Ymedian:	21.000000000
Y@Xmin:	67.000000000	Y@Xmax:	32.000000000	Y@Xrange:	35.000000000

Date	Time	File Source
Feb 20. 2014	4:51:26 PM	CLIPBRD.PRN

Artemisia plant particle chronosere

Rank 1 Eqn 8160 [Line Robust None, Gaussian Errors] y=a+bx

r^2=0.44574765 DF Adj r^2=0.43722069 FitStdErr=28.718286 Fstat=105.35443

a=8.7291173

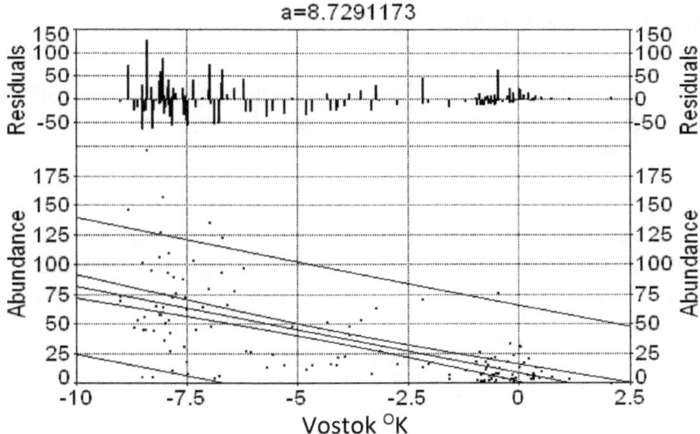

Figure 5D2. Linear regression of the Artemisia abundance chronosere with respect to Vostok temperature differences. Numeric table:

Rank 1 Eqn 8160 [Line Robust None, Gaussian Errors] y=a+bx

r^2 Coef Det	DF Adj r^2	Fit Std Err	F-value
0.4457476488	0.4372206895	28.718286312	105.35443262

Parm	Value	Std Error	t-value	95% Confidence Limits		P>|t|
a	8.729117279	3.739825408	2.334097538	1.330850524	16.12738403	0.02111
b	-7.33690462	0.714803165	-10.2642307	-8.75095579	-5.92285345	0.00000

Area Xmin-Xmax	Area Precision		
379.61792250	0.0000000000		
Function min	X-Value	Function max	X-Value
-6.384906236	2.0600000000	74.907870077	-9.019982709
1st Deriv min	X-Value	1st Deriv max	X-Value
-7.336904619	-8.783728481	-7.336904619	-3.479993395
2nd Deriv min	X-Value	2nd Deriv max	X-Value
-7.06514e-09	-3.893677940	7.065143e-09	-1.264059258

Procedure	Minimization	Iterations	
LevMarqdt	Least Squares	6	
r^2 Coef Det	DF Adj r^2	Fit Std Err	r^2 Attainable
0.4457476488	0.4372206895	28.718286312	0.9605296126

Source	Sum of Squares	DF	Mean Square	F Statistic	P>F
Regr	86890.011	1	86890.011	105.354	0.00000
Error	108040.94	131	824.73997		
Total	194930.95	132			
Lack Fit	100346.94	119	843.25156	1.31518	0.31036
Pure Err	7694	12	641.16667		

Description: Artemisia plant particle chronosere

X Variable: Vostok OK

Xmin:	-9.020000000	Xmax:	2.0600000000	Xrange:	11.080000000
Xmean:	-3.903458647	Xstd:	3.4969151005	Xmedian:	-3.850000000
X@Ymin:	2.0600000000	X@Ymax:	-8.410000000	X@Yrange:	10.470000000

Y Variable: Abundance

Ymin:	0.0000000000	Ymax:	198.00000000	Yrange:	198.00000000
Ymean:	37.368421053	Ystd:	38.428499857	Ymedian:	21.000000000
Y@Xmin:	74.000000000	Y@Xmax:	0.0000000000	Y@Xrange:	74.000000000

Date	Time	File Source
Feb 19, 2014	12:33:34 PM	CLIPBRD.PRN

Poaceae

Poaceae plant particle chronosere

Rank 101 Eqn 8004 [Lorentzian] $y=a+b/(1+((x-c)/d)^2)$

r^2=0.45203207 DF Adj r^2=0.43490807 FitStdErr=41.364824 Fstat=35.471745

a=29.353344 b=133.94265

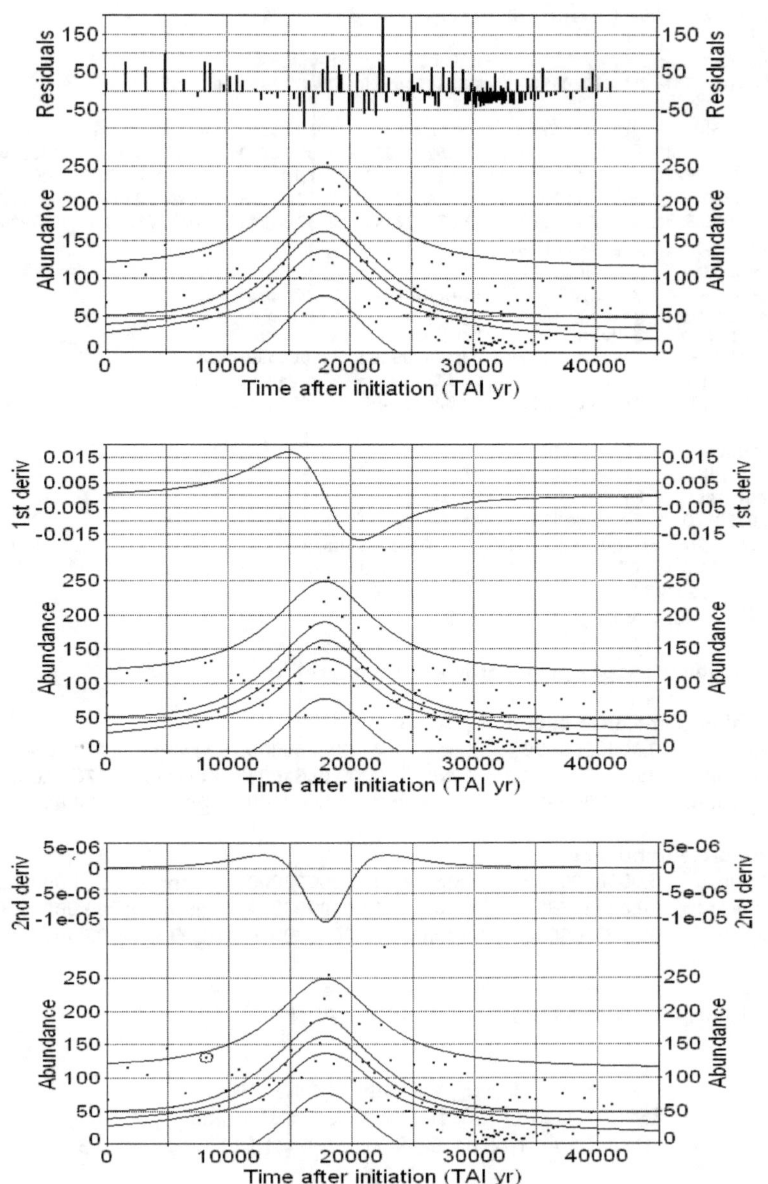

Figure 5E1. The Cauchy-Lorentz trend line fitted to the Poaceae oscillogram of Figure 5A. Graph types are same as in Figure 5B. The numeric table:

Rank 101 Eqn 8004 [Lorentzian] $y=a+b/(1+((x-c)/d)^2)$

r^2 Coef Det	DF Adj r^2	Fit Std Err	F-value
0.4520320670	0.4349080691	41.364824411	35.471745169

Parm	Value	Std Error	t-value	95% Confidence Limits		P>|t|
a	29.35334358	8.506375633	3.450746223	12.52327105	46.18341611	0.00076
b	133.9426494	13.25456703	10.10539606	107.7181639	160.1671349	0.00000
c	17908.53292	471.9058326	37.94937820	16974.85567	18842.21016	0.00000
d	5022.815126	1030.557363	4.873882140	2983.832142	7061.798109	0.00000

Area Xmin-Xmax	Area Precision		
2993874.5361	2.686463e-14		
Function min	X-Value	Function max	X-Value
35.335947874	41138.000000	163.29599295	17908.532038
1st Deriv min	X-Value	1st Deriv max	X-Value
-0.017320626	20808.448675	0.0173206261	15008.608334
2nd Deriv min	X-Value	2nd Deriv max	X-Value
-1.06183e-05	17908.523866	2.654572e-06	22931.367722

Procedure	Minimization	Iterations
LevMarqdt	Least Squares	17
r^2 Coef Det	DF Adj r^2	Fit Std Err
0.4520320670	0.4349080691	41.364824411

Source	Sum of Squares	DF	Mean Square	F Statistic	P>F
Regr	182081.65	3	60693.883	35.4717	0.00000
Error	220725.28	129	1711.0487		
Total	402806.93	132			

Description: Poaceae plant particle chronosere

X Variable: Time after initiation (TAI yr)

Xmin:	0.0000000000	Xmax:	41138.000000	Xrange:	41138.000000
Xmean:	25563.609023	Xstd:	9209.0461001	Xmedian:	27522.000000
X@Ymin:	30252.000000	X@Ymax:	22611.000000	X@Yrange:	7641.0000000

Y Variable: Abundance

Ymin:	5.0000000000	Ymax:	296.00000000	Yrange:	291.00000000
Ymean:	70.451127820	Ystd:	55.240996272	Ymedian:	60.000000000
Y@Xmin:	69.000000000	Y@Xmax:	60.000000000	Y@Xrange:	9.0000000000

Date	Time	File Source
Feb 20, 2014	4:39:16 PM	CLIPBRD.PRN

Poaceae plant particles chronosere

Rank 1 Eqn 8160 [Line Robust None, Gaussian Errors] $y=a+bx$

r^2=0.45211588 DF Adj r^2=0.4436869 FitStdErr=41.044708 Fstat=108.10166

a=28.989067

b=-10.621878

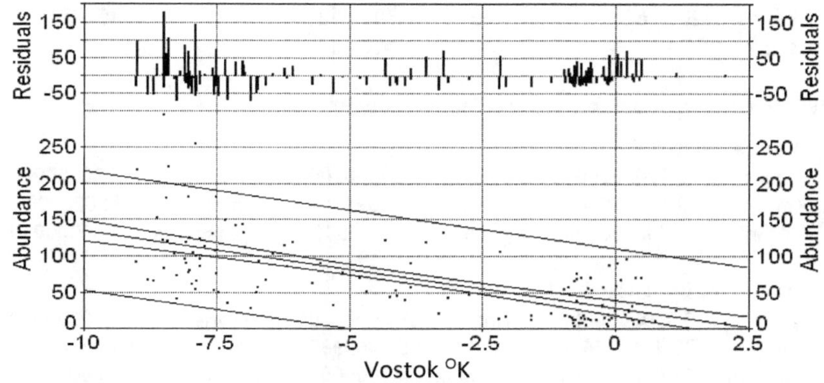

Figure 5E2. Linear regression of the Poaceae abundance chronosere with respect to Vostok temperature differences. Numeric table:

Rank 1 Eqn 8160 [Line Robust None, Gaussian Errors] y=a+bx

r^2 Coef Det	DF Adj r^2	Fit Std Err	F-value
0.4521158840	0.4436868976	41.044708452	108.10165703

Parm	Value	Std Error	t-value	95% Confidence Limits		P>|t|
a	28.98906747	5.345027968	5.423557677	18.41532715	39.56280780	0.00000
b	-10.6218777	1.021609966	-10.3971947	-12.6428659	-8.60088949	0.00000

Area Xmin-Xmax	Area Precision		
730.76147666	7.596343e-20		
Function min	X-Value	Function max	X-Value
7.1079994121	2.0600000000	124.79822065	-9.019982709
1st Deriv min	X-Value	1st Deriv max	X-Value
-10.62187770	-8.530530557	-10.62187770	-1.048470385
2nd Deriv min	X-Value	2nd Deriv max	X-Value
-2.82606e-08	-4.685859187	1.413029e-08	-3.510873359

Procedure	Minimization	Iterations	
LevMarqdt	Least Squares	6	
r^2 Coef Det	DF Adj r^2	Fit Std Err	r^2 Attainable
0.4521158840	0.4436868976	41.044708452	0.9551546943

Source	Sum of Squares	DF	Mean Square	F Statistic	P>F
Regr	182115.41	1	182115.41	108.102	0.00000
Error	220691.52	131	1684.6681		
Total	402806.93	132			

Lack Fit	202627.52	119	1702.7523	1.13115
Pure Err	18064	12	1505.3333	

Description: Poaceae plant particles chronosere

X Variable: Vostok OK

Xmin:	-9.020000000	Xmax:	2.0600000000	Xrange:	11.080000000
Xmean:	-3.903458647	Xstd:	3.4969151005	Xmedian:	-3.850000000
X@Ymin:	-0.150000000	X@Ymax:	-8.500000000	X@Yrange:	8.3500000000

Y Variable: Abundance

Ymin:	5.0000000000	Ymax:	296.00000000	Yrange:	291.00000000
Ymean:	70.451127820	Ystd:	55.240996272	Ymedian:	60.000000000
Y@Xmin:	93.000000000	Y@Xmax:	10.000000000	Y@Xrange:	83.000000000

Date	Time	File Source
Feb 19, 2014	2:13:00 PM	CLIPBRD.PRN

Picea

Picea Plant Particle chronosere
Rank 42 Eqn 6059 High Precision Polynomial Order 9
r^2=0.39193331 DF Adj r^2=0.34209177 FitStdErr=29.369316 Fstat=8.8089381
a=24.148068 b=-5.2111714e-06 c=7.7736517e-07 d=1.6555506e-11 e=-6.3871795e-14
f=1.0103894e-17 q=-6.7080093e-22 h=2.236423e-26 i=-3.6969915e-31 j=2.4143382e-36

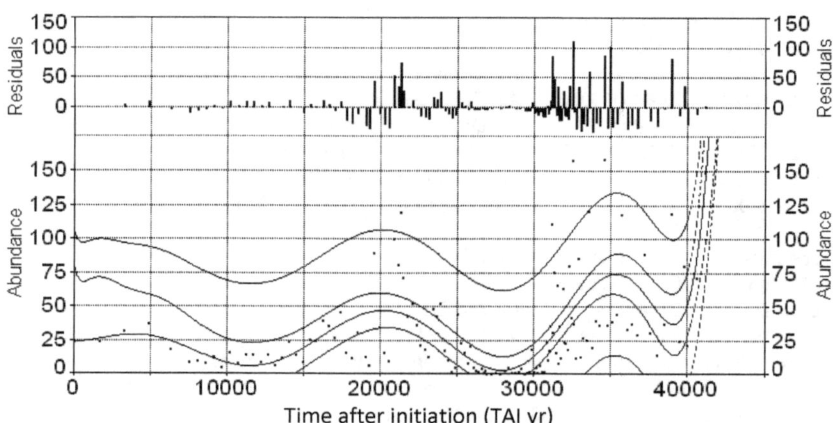

Figure 5F1. Order 9 polynomial trend line fitted to the Picea oscillogram of Figure 5A. Some details in the caption of Figure 5B apply. Numeric table:

Rank 42 Eqn 6059 High Precision Polynomial Order 9

r^2 Coef Det	DF Adj r^2	Fit Std Err	F-value
0.3919333068	0.3420917745	29.369316431	8.8089380963

Parm	Value	Std Error	t-value	95% Confidence Limits		P>|t|
a	24.14806799	28.97684356	0.833357434	-33.2098171	81.50595312	0.40626
b	-5.2112e-06	0.043693523	-0.00011927	-0.08649386	0.086483439	0.99991
c	7.77365e-07	2.0645e-05	0.037653828	-4.0088e-05	4.1643e-05	0.97002
d	1.65555e-11	4.25689e-09	0.003889113	-8.4097e-09	8.4428e-09	0.99690
e	-6.3872e-14	4.68159e-13	-0.13643177	-9.9056e-13	8.62821e-13	0.89170
f	1.01039e-17	3.00669e-17	0.336047148	-4.9412e-17	6.96195e-17	0.73741
g	-6.708e-22	1.16461e-21	-0.57598849	-2.9761e-21	1.63447e-21	0.56568
h	2.23642e-26	2.67907e-26	0.834776590	-3.0666e-26	7.53947e-26	0.40546
i	-3.697e-31	3.36994e-31	-1.09704818	-1.0368e-30	2.97361e-31	0.27476
j	2.41434e-36	1.78483e-36	1.352700562	-1.1186e-36	5.9473e-36	0.17863

Area Xmin-Xmax	Area Precision		
1293057.0489	1.261666e-16		
Function min	**X-Value**	**Function max**	**X-Value**
2.8211116932	27968.781432	74.192532398	35355.450878
1st Deriv min	**X-Value**	**1st Deriv max**	**X-Value**
-0.008912438	24413.981886	0.0153719126	32205.100867
2nd Deriv min	**X-Value**	**2nd Deriv max**	**X-Value**
-9.12126e-06	35725.594968	0.0001245116	41138.000000

Description: Picea plant particles chronosere

X Variable: Time after initiation (TAI

Xmin:	0.0000000000	Xmax:	41138.000000	Xrange:	41138.000000
Xmean:	25563.609023	Xstd:	9209.0461001	Xmedian:	27522.000000
X@Ymin:	29366.000000	X@Ymax:	34897.000000	X@Yrange:	5531.0000000

Y Variable: Abundance

Ymin:	0.0000000000	Ymax:	175.00000000	Yrange:	175.00000000
Ymean:	32.165413534	Ystd:	36.356646771	Ymedian:	20.000000000
Y@Xmin:	24.000000000	Y@Xmax:	149.00000000	Y@Xrange:	125.00000000

Date	Time	File Source
Feb 19, 2014	3:05:42 PM	CLIPBRD.PRN

Picea Plant Particle chronosere

Rank 82 Eqn 6723 $y=a+bx^2+cx^4+dx^6+ex^8$

r^2=0.16620408 DF Adj r^2=0.13311694 FitStdErr=33.792851 Fstat=6.3288623

a=48.321443 b=-7.2192486 c=0.32838777

d=-0.0052054149 e=2.7583345e-05

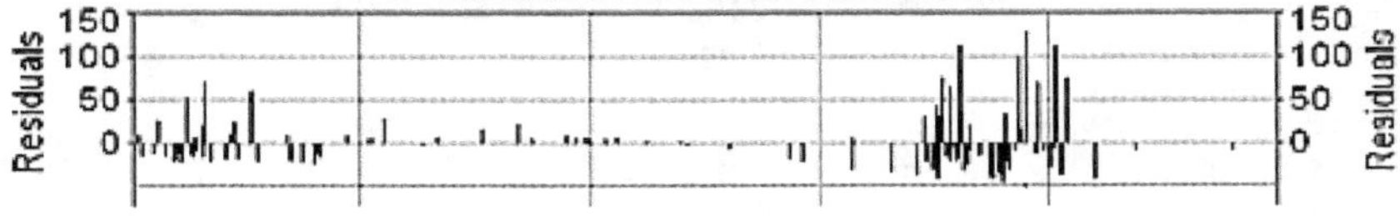

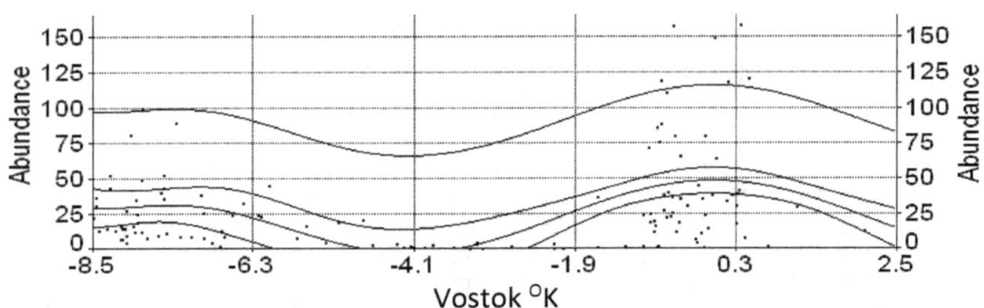

Figure 5F2. Mixed order polynomial fitted the Poaceae abundance chronosere with respect to Vostok temperature differences. Numeric table:

Rank 82 Eqn 6723 $y=a+bx^2+cx^4+dx^6+ex^8$

r^2 Coef Det	DF Adj r^2	Fit Std Err	F-value
0.1662040808	0.1331169411	33.792850937	6.3288623058

Parm	Value	Std Error	t-value	95% Confidence Limits		P>|t|
a	48.32144281	4.719288505	10.23913727	38.98282253	57.66006310	0.00000
b	-7.21924861	1.869329858	-3.86194474	-10.9183150	-3.52018216	0.00018
c	0.328387770	0.115948609	2.832183798	0.098946399	0.557829142	0.00538
d	-0.00520541	0.002328142	-2.23586701	-0.00981239	-0.00059844	0.02711
e	2.75833e-05	1.47882e-05	1.865224256	-1.6799e-06	5.68466e-05	0.06446

Area Xmin-Xmax	Area Precision		
271.03675708	4.096208e-19		
Function min	X-Value	Function max	X-Value
-2.773558985	-4.136891184	48.321442813	3.867755e-10
1st Deriv min	X-Value	1st Deriv max	X-Value
-19.38436867	2.0600000000	19.391925552	-2.096841304
2nd Deriv min	X-Value	2nd Deriv max	X-Value
-14.43849721	1.029148e-05	104.35798930	-9.019982709

Soln Vector	Covar Matrix		
GaussElim	LUDecomp		
r^2 Coef Det	DF Adj r^2	Fit Std Err	r^2 Attainable
0.1662040808	0.1331169411	33.792850937	0.9551218520

Source	Sum of Squares	DF	Mean Square	F Statistic	P>F
Regr	28909.149	4	7227.2872	6.32886	0.00011
Error	145028.51	127	1141.9568		
Total	173937.66	131			
Lack Fit	137222.51	115	1193.2392	1.83434	0.11925
Pure Err	7806	12	650.5		

Description: Picea plant particles spectrum

X Variable: Vostok OK

Xmin:	-9.020000000	Xmax:	2.0600000000	Xrange:	11.080000000
Xmean:	-3.878787879	Xstd:	3.4985988537	Xmedian:	-3.710000000
X@Ymin:	-3.340000000	X@Ymax:	0.0900000000	X@Yrange:	3.4300000000

Y Variable: Abundance

Ymin:	0.0000000000	Ymax:	175.00000000	Yrange:	175.00000000
Ymean:	32.340909091	Ystd:	36.438556366	Ymedian:	20.500000000
Y@Xmin:	24.000000000	Y@Xmax:	13.000000000	Y@Xrange:	11.000000000

Date	Time	File Source
Feb 19, 2014	5:54:46 PM	CLIPBRD.PRN

Betula

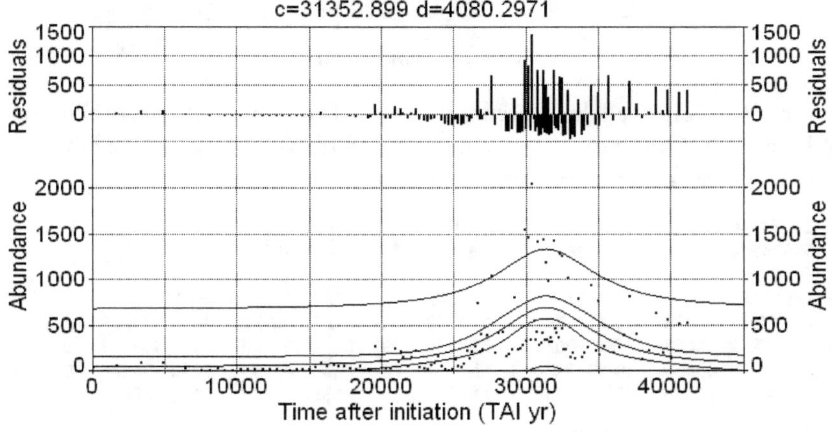

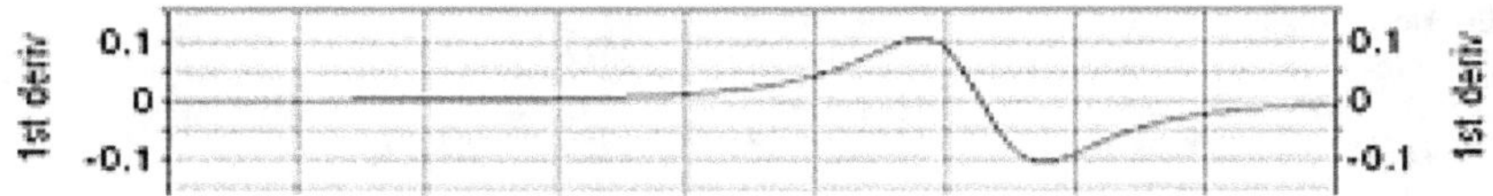

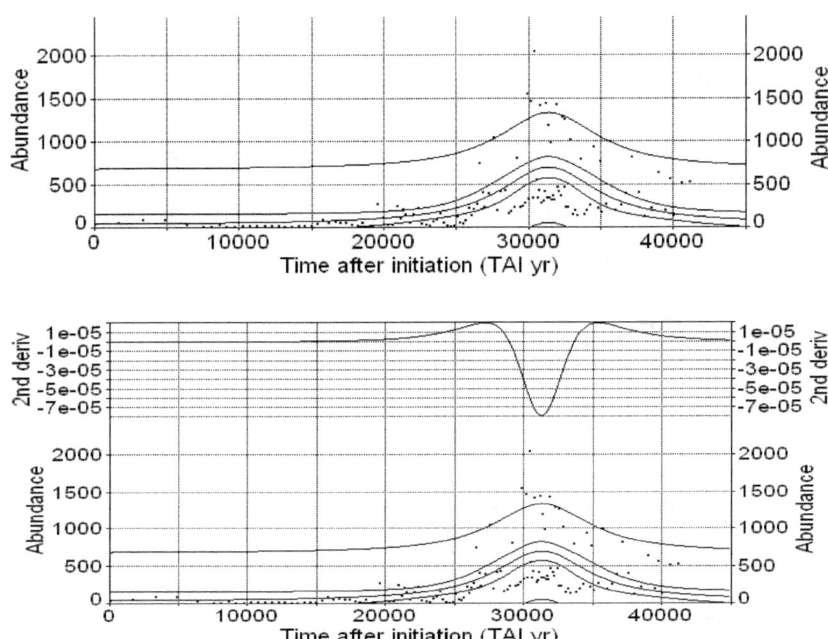

Figure 5G1. The Cauchy-Lorentz trend line fitted to the Betula oscillogram of Figure 5A. Graph types are same as in Figure 5B. The numeric table:

Rank 91 Eqn 8004 [Lorentzian] $y=a+b/(1+((x-c)/d)^2)$

r^2 Coef Det	DF Adj r^2	Fit Std Err	F-value
0.3619249161	0.3419850697	315.99187903	24.390188214

Parm	Value	Std Error	t-value	95% Confidence Limits		P>|t|
a	38.72103891	61.44247618	0.630199844	-82.8444051	160.2864829	0.52968
b	658.2220024	77.33600720	8.511197129	505.2108181	811.2331868	0.00000
c	31352.89868	497.0537789	63.07747775	30369.46560	32336.33176	0.00000
d	4080.297069	1031.572163	3.955416030	2039.306279	6121.287859	0.00013

Area Xmin-Xmax	Area Precision		
8621791.2389	2.984255e-13		
Function min	X-Value	Function max	X-Value
49.683464377	1.733943e-10	696.94304132	31352.898786
1st Deriv min	X-Value	1st Deriv max	X-Value
-0.104778580	33708.664227	0.1042413514	28796.600000

2nd Deriv min	X-Value	2nd Deriv max	X-Value
-7.90713e-05	31352.870676	1.976782e-05	27272.574404

Procedure	Minimization	Iterations
LevMarqdt	Least Squares	18
r^2 Coef Det	DF Adj r^2	Fit Std Err
0.3619249161	0.3419850697	315.99187903

Source	Sum of Squares	DF	Mean Square	F Statistic	P>F
Regr	7306144.4	3	2435381.5	24.3902	0.00000
Error	12880762	129	99850.868		
Total	20186906	132			

Description: Betula plant particle s chronosere

X Variable: Time after initiation (TAI yr)

Xmin:	0.0000000000	Xmax:	41138.000000	Xrange:	41138.000000
Xmean:	25563.609023	Xstd:	9209.0461001	Xmedian:	27522.000000
X@Ymin:	24593.000000	X@Ymax:	30370.000000	X@Yrange:	5777.0000000

Y Variable: Abundance

Ymin:	14.000000000	Ymax:	2044.0000000	Yrange:	2030.0000000
Ymean:	331.85714286	Ystd:	391.06407176	Ymedian:	216.00000000
Y@Xmin:	55.000000000	Y@Xmax:	532.00000000	Y@Xrange:	477.00000000

Date	Time	File Source
Feb 19, 2014	6:07:36 PM	CLIPBRD.PRN

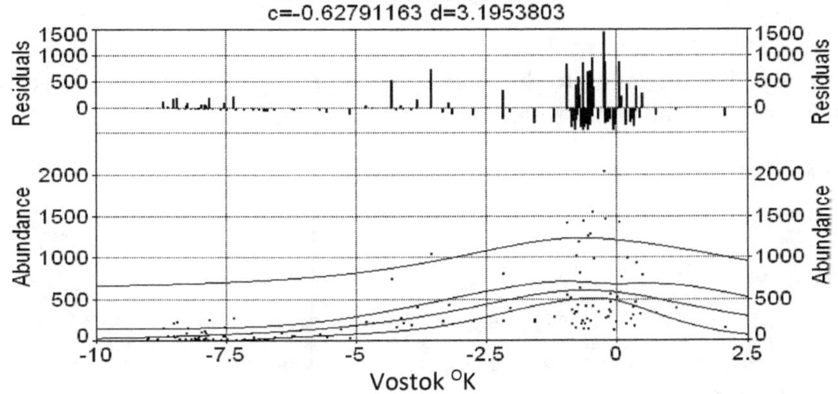

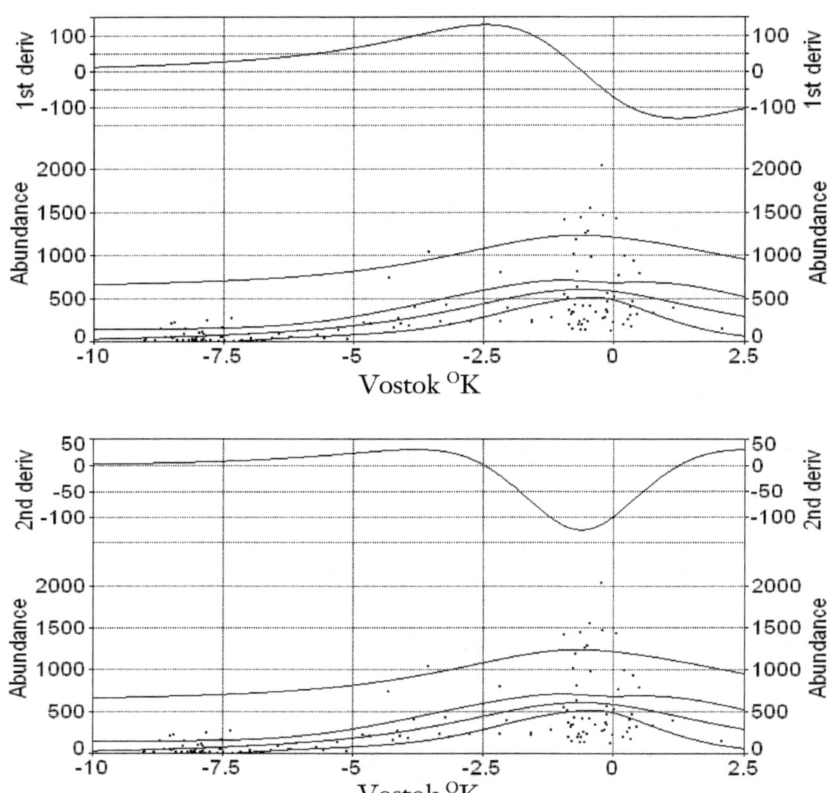

Figure 5G2. The Cauchy-Lorentz trend line fitted to the Betula oscillogram of Figure 5A in relation to the Vostok chronosere. Graph types are same as in Figure 5B. The numeric table:

Rank 101 Eqn 8004 [Lorentzian] $y=a+b/(1+((x-c)/d)^2)$

r^2 Coef Det	DF Adj r^2	Fit Std Err	F-value
0.3723486132	0.3527345073	313.40020389	25.509368262

Parm	Value	Std Error	t-value	95% Confidence Limits		P>|t|
a	-31.8638105	106.2624840	-0.29985945	-242.106738	178.3791166	0.76477
b	638.8916553	103.4312046	6.176972003	434.2504837	843.5328269	0.00000
c	-0.62791163	0.545389367	-1.15130890	-1.70697785	0.451154586	0.25173
d	3.195380290	1.301758561	2.454664319	0.619819095	5.770941485	0.01543

Area Xmin-Xmax	Area Precision
3538.7449165	5.087482e-09

Function min	X-Value	Function max	X-Value
49.033908624	-9.019982709	607.02784487	-0.627911752
1st Deriv min	X-Value	1st Deriv max	X-Value
-129.8663274	1.2169419022	129.86632738	-2.472764586
2nd Deriv min	X-Value	2nd Deriv max	X-Value
-125.1445980	-0.627916617	31.286149589	-3.823264802

Procedure	Minimization	Iterations	
LevMarqdt	Least Squares	18	
r^2 Coef Det	DF Adj r^2	Fit Std Err	r^2 Attainable
0.3723486132	0.3527345073	313.40020389	0.9559784700

Source	Sum of Squares	DF	Mean Square	F Statistic	P>F
Regr	7516566.6	3	2505522.2	25.5094	0.00000
Error	12670340	129	98219.688		
Total	20186906	132			
Lack Fit	11781681	117	100698.13	1.35978	0.28566
Pure Err	888658.5	12	74054.875		

Description: Betula plant particle distribution

X Variable: Time after initiation (TAI yr)

Xmin:	-9.020000000	Xmax:	2.0600000000	Xrange:	11.080000000
Xmean:	-3.903458647	Xstd:	3.4969151005	Xmedian:	-3.850000000
X@Ymin:	-6.770000000	X@Ymax:	-0.260000000	X@Yrange:	6.5100000000

Y Variable: Abundance

Ymin:	14.000000000	Ymax:	2044.0000000	Yrange:	2030.0000000
Ymean:	331.85714286	Ystd:	391.06407176	Ymedian:	216.00000000
Y@Xmin:	33.000000000	Y@Xmax:	160.00000000	Y@Xrange:	127.00000000

Date	Time	File Source
Feb 21, 2014	5:53:57 AM	CLIPBRD.PRN

Alnus

Alnus plant particle chronosere

Rank 122 Eqn 8004 [Lorentzian] $y=a+b/(1+((x-c)/d)^2)$

r^2=0.37022687 DF Adj r^2=0.35054646 FitStdErr=128.08105 Fstat=25.278556

a=3.3593313 b=361.403

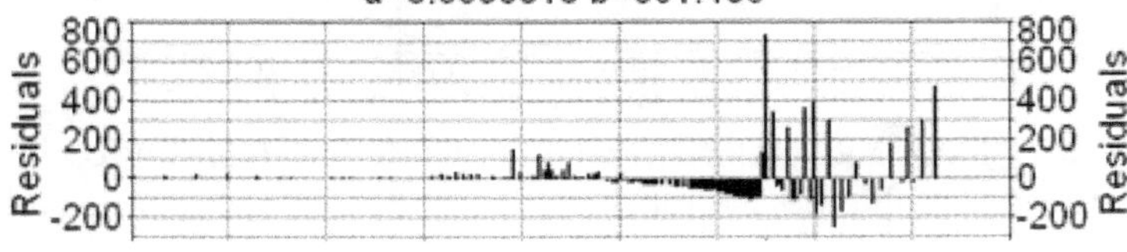

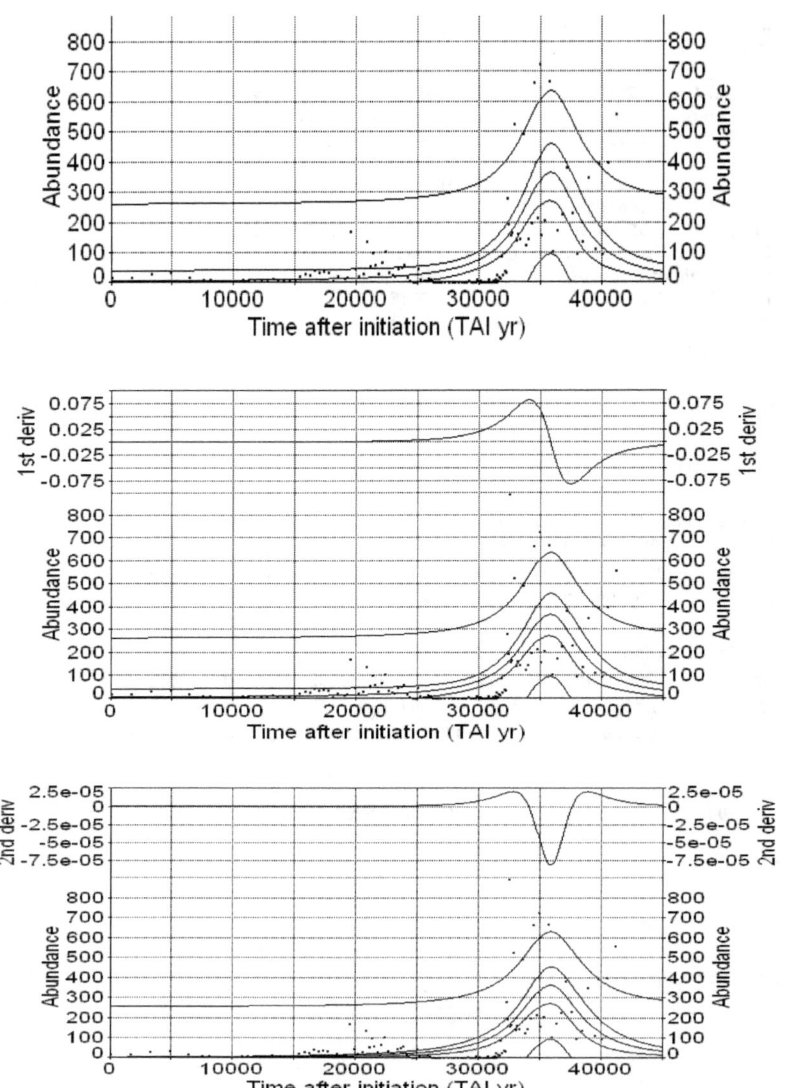

Figure 5H1. The Cauchy-Lorentz trend line fitted to the Alnus oscillogram of Figure 5A. Graph types are same as in Figure 5B. The numeric table:

Rank 122 Eqn 8004 [Lorentzian] $y=a+b/(1+((x-c)/d)^2)$

r^2 Coef Det	DF Adj r^2	Fit Std Err	F-value
0.3702268678	0.3505464574	128.08105437	25.278555881

Parm	Value	Std Error	t-value	95% Confidence Limits		P>|t|
a	3.359331293	18.12553249	0.185336971	-32.5024787	39.22114126	0.85326
b	361.4030014	47.64497988	7.585332229	267.1362418	455.6697610	0.00000
c	35787.33349	379.8487307	94.21469809	35035.79347	36538.87350	0.00000
d	2855.439588	614.8335852	4.644247902	1638.976281	4071.902895	0.00001

Area Xmin-Xmax	Area Precision		
2792161.2174	7.46085e-12		
Function min	X-Value	Function max	X-Value
5.6455771985	1.733943e-10	364.76233272	35787.333606
1st Deriv min	X-Value	1st Deriv max	X-Value
-0.082207355	37435.930801	0.0822073547	34138.747746
2nd Deriv min	X-Value	2nd Deriv max	X-Value
-8.86494e-05	35787.338746	2.216235e-05	38642.773170

Procedure	Minimization	Iterations
LevMarqdt	Least Squares	33
r^2 Coef Det	DF Adj r^2	Fit Std Err
0.3702268678	0.3505464574	128.08105437

Source	Sum of Squares	DF	Mean Square	F Statistic	P>F
Regr	1244065.7	3	414688.55	25.2786	0.00000
Error	2116213.6	129	16404.756		
Total	3360279.2	132			

Description: Alnus plant particle chronosere

X Variable: Time after initio

Xmin:	0.0000000000	Xmax:	41138.000000	Xrange:	41138.000000
Xmean:	25563.609023	Xstd:	9209.0461001	Xmedian:	27522.000000
X@Ymin:	30252.000000	X@Ymax:	32479.000000	X@Yrange:	2227.0000000

Y Variable: Abundance

Ymin:	0.0000000000	Ymax:	895.00000000	Yrange:	895.00000000
Ymean:	87.075187970	Ystd:	159.55143676	Ymedian:	19.000000000
Y@Xmin:	19.000000000	Y@Xmax:	559.00000000	Y@Xrange:	540.00000000

Date	Time	File Source
Feb 20, 2014	9:42:36 AM	CLIPBRD.PRN

Alnus plant particle chronosere

Rank 1 Eqn 8160 [Line Robust None, Gaussian Errors] $y=a+bx$

r^2=0.16357886 DF Adj r^2=0.15071085 FitStdErr=146.47535 Fstat=25.619667

a=159.10774

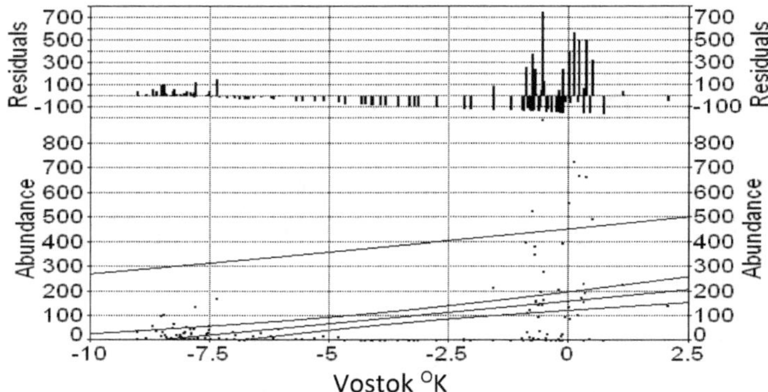

Figure 5H2. Linear trend line fitted to the Alnus oscillogram of Figure 5A in relation to the Vostok chronosere. Numeric table:

Rank 1 Eqn 8160 [Line Robust None, Gaussian Errors] y=a+bx

r^2 Coef Det	DF Adj r^2	Fit Std Err	F-value
0.1635788641	0.1507108466	146.47534732	25.619667268

Parm	Value	Std Error	t-value	95% Confidence Limits		P>|t|
a	159.1077410	19.07468362	8.341304327	121.3734667	196.8420153	0.00000
b	18.45352022	3.645796993	5.061587426	11.24126410	25.66577634	0.00000

Lack Fit	2710935.6	119	22780.971	2.74269	0.02674
Pure Err	99673	12	8306.0833		

Description: Alnus plant particle chronosere

X Variable: Vostok oH

Xmin:	-9.020000000	Xmax:	2.0600000000	Xrange:	11.080000000
Xmean:	-3.903458647	Xstd:	3.4969151005	Xmedian:	-3.850000000
X@Ymin:	-0.150000000	X@Ymax:	-0.560000000	X@Yrange:	0.4100000000

Y Variable: Abundance

Ymin:	0.0000000000	Ymax:	895.00000000	Yrange:	895.00000000
Ymean:	87.075187970	Ystd:	159.55143676	Ymedian:	19.000000000
Y@Xmin:	5.0000000000	Y@Xmax:	143.00000000	Y@Xrange:	138.00000000

Date	Time	File Source
Feb 20, 2014	8:47:42 AM	CLIPBRD.PRN

Sphagnum

Sphagnum plant particle chronosere
Rank 123 Eqn 8004 [Lorentzian] $y=a+b/(1+((x-c)/d)^2)$
r^2=0.2517377 DF Adj r^2=0.22835451 FitStdErr=17.472265 Fstat=14.46648
a=0.53661399 b=31.739859

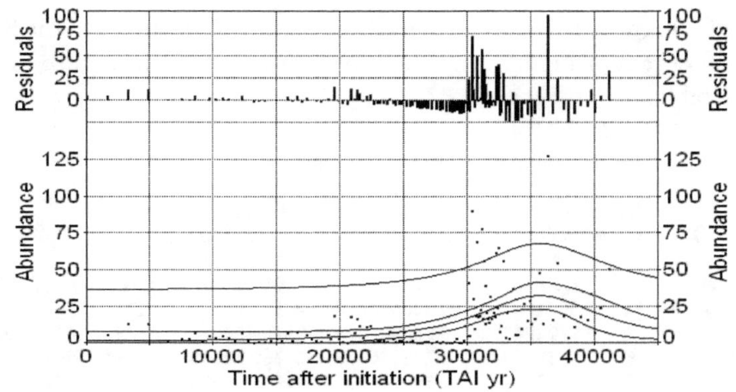
Residuals
100
75
50
25
0
Abundance
125
100
75
50
25
0
0
10000
20000
30000
40000
Time after initiation (TAI yr)

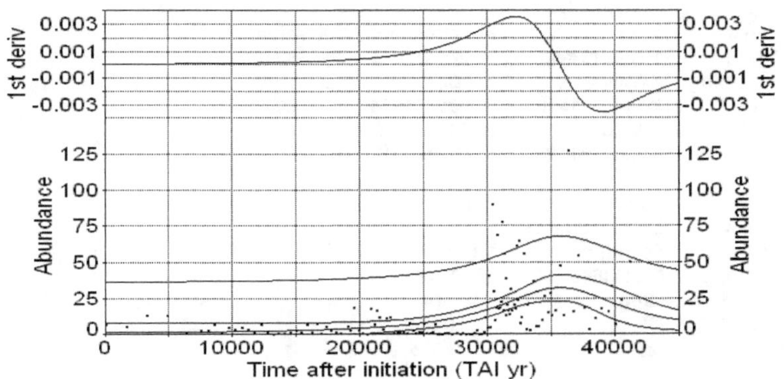
1st deriv
0.003
0.001
-0.001
-0.003
Abundance
125
100
75
50
25
0
0
10000
20000
30000
40000
Time after initiation (TAI yr)

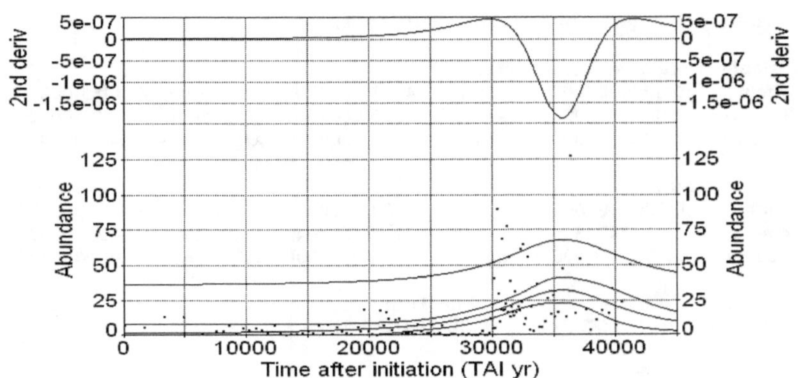
2nd deriv
5e-07
0
-5e-07
-1e-06
-1.5e-06
Abundance
125
100
75
50
25
0
0
10000
20000
30000
40000
Time after initiation (TAI yr)

Figure 5I1. Figure 5H1. The Cauchy-Lorentz trend line fitted to the Sphagnum oscillogram of Figure 5A. Graph types are same as in Figure 5B. The numeric table:

Rank 123 Eqn 8004 [Lorentzian] $y=a+b/(1+((x-c)/d)^2)$

r^2 Coef Det	DF Adj r^2	Fit Std Err	F-value
0.2517377022	0.2283545054	17.472264510	14.466479507

Parm	Value	Std Error	t-value	95% Confidence Limits		P>|t|
a	0.536613991	3.564274704	0.150553489	-6.51539081	7.588618789	0.88056
b	31.73985909	4.970198093	6.386035023	21.90620043	41.57351775	0.00000
c	35683.99456	952.3566149	37.46915179	33799.73368	37568.25545	0.00000
d	5822.045212	1976.288711	2.945948728	1911.909595	9732.180830	0.00382

Area Xmin-Xmax	Area Precision		
421562.90671	1.876811e-11		
Function min	X-Value	Function max	X-Value
1.3596135291	1.733943e-10	32.276473082	35683.987809
1st Deriv min	X-Value	1st Deriv max	X-Value
-0.003540962	39045.351508	0.0035409624	32322.639656
2nd Deriv min	X-Value	2nd Deriv max	X-Value
-1.87277e-06	35683.994392	4.681919e-07	29861.934863

Procedure	Minimization	Iterations
LevMarqdt	Least Squares	17
r^2 Coef Det	DF Adj r^2	Fit Std Err
0.2517377022	0.2283545054	17.472264510

Source	Sum of Squares	DF	Mean Square	F Statistic	P>F
Regr	13248.982	3	4416.3273	14.4665	0.00000
Error	39381.123	129	305.28003		
Total	52630.105	132			

Description: Sphagnum plant particle chronosere

X Variable: Time after initiation (TAI yr)

Xmin:	0.0000000000	Xmax:	41138.000000	Xrange:	41138.000000
Xmean:	25563.609023	Xstd:	9209.0461001	Xmedian:	27522.000000
X@Ymin:	26394.000000	X@Ymax:	36360.000000	X@Yrange:	9966.0000000

Y Variable: Abundance

Ymin:	0.0000000000	Ymax:	128.00000000	Yrange:	128.00000000
Ymean:	13.789473684	Ystd:	19.967797041	Ymedian:	6.0000000000
Y@Xmin:	7.0000000000	Y@Xmax:	51.000000000	Y@Xrange:	44.000000000

Date	Time	File Source
Feb 20, 2014	1:07:32 PM	CLIPBRD.PRN

Sphagnum plant particle chronosere
Rank 1 Eqn 8160 [Line Robust None, Gaussian Errors] y=a+bx
r^2=0.24132509 DF Adj r^2=0.22965317 FitStdErr=17.458597 Fstat=41.669478
a=24.739014

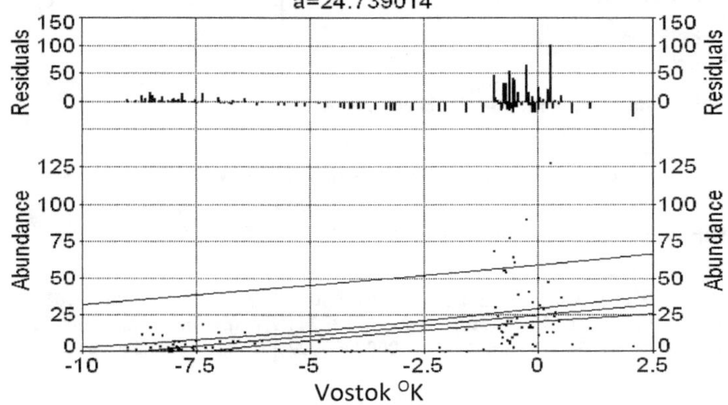

Figure 5I2. Linear trend line fitted to the Sphagnum oscillogram of Figure 5A in relation to the Vostok chronosere. Numeric table:

Rank 1 Eqn 8160 [Line Robust None, Gaussian Errors] y=a+bx

r^2 Coef Det	DF Adj r^2	Fit Std Err	F-value
0.2413250942	0.2296531725	17.458596639	41.669478051

Parm	Value	Std Error	t-value	95% Confidence Limits		P>|t|
a	24.73901408	2.273537585	10.88128661	20.24141441	29.23661375	0.00000
b	2.805086819	0.434547521	6.455190009	1.945448164	3.664725473	0.00000

Parm	Value	Std Error	t-value	95% Confidence Limits		P>|t|
a	24.73901408	2.273537585	10.88128661	20.24141441	29.23661375	0.00000
b	2.805086819	0.434547521	6.455190009	1.945448164	3.664725473	0.00000

Area Xmin-Xmax	Area Precision		
165.94861643	8.362702e-20		
Function min	X-Value	Function max	X-Value
-0.562820521	-9.019982709	30.517492928	2.0600000000
1st Deriv min	X-Value	1st Deriv max	X-Value
2.8050868188	-2.865880377	2.8050868188	-9.019982709
2nd Deriv min	X-Value	2nd Deriv max	X-Value
-7.06513e-09	-3.017999514	3.532562e-09	-1.755504635

Procedure	Minimization	Iterations	
LevMarqdt	Least Squares	7	
r^2 Coef Det	DF Adj r^2	Fit Std Err	r^2 Attainable
0.2413250942	0.2296531725	17.458596639	0.9337071438

Source	Sum of Squares	DF	Mean Square	F Statistic	P>F
Regr	12700.965	1	12700.965	41.6695	0.00000
Error	39929.14	131	304.8026		
Total	52630.105	132			
Lack Fit	36440.14	119	306.21967	1.05321	0.49915
Pure Err	3489	12	290.75		

Description: Sphagnum plant particle chronosere

X Variable: Vostok oK

Xmin:	-9.020000000	Xmax:	2.0600000000	Xrange:	11.080000000
Xmean:	-3.903458647	Xstd:	3.4969151005	Xmedian:	-3.850000000
X@Ymin:	-4.140000000	X@Ymax:	0.2500000000	X@Yrange:	4.3900000000

Y Variable: Abundance

Ymin:	0.0000000000	Ymax:	128.00000000	Yrange:	128.00000000
Ymean:	13.789473684	Ystd:	19.967797041	Ymedian:	6.0000000000
Y@Xmin:	1.0000000000	Y@Xmax:	4.0000000000	Y@Xrange:	3.0000000000

Date	Time	File Source
Feb 20, 2014	1:13:11 PM	CLIPBRD.PRN

T and n oscillations

T, the total resonator abundance, and n, the number of resonators within complexes. In this case the complexes are paleorelevés and the resonators are the palynomorph taxa. The Cwynar chronosere has 133 paleorelevés and one paleorelevé can have at most 93 palynomorph taxa. The oscillograms and fitted regression lines are in Figure 6.

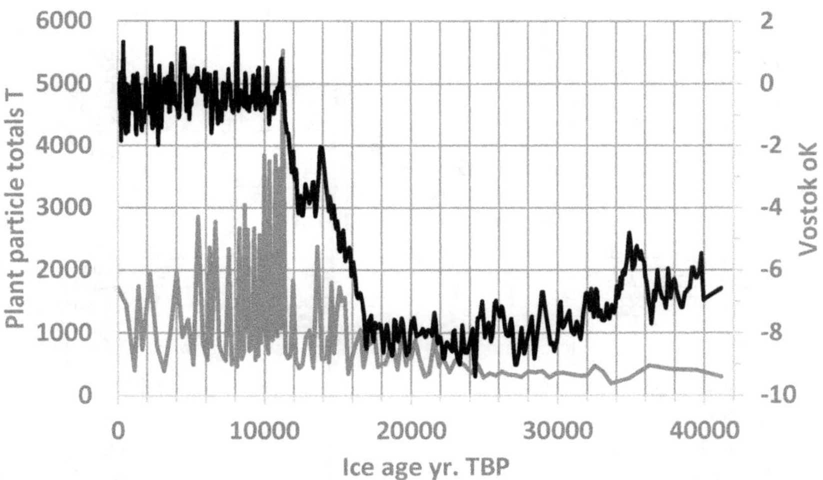

Figure 6A. Plant particle count totals T in 133 paleorelevés arranged along the time axis scaled as TBP. The shortened Vostok temperature chronosere (637 points) is included for quick reference. The particle counts time scale is Cwynar's carbon age scale. Conversion from TBP to TAI is TAI=TBP-41138 yr.

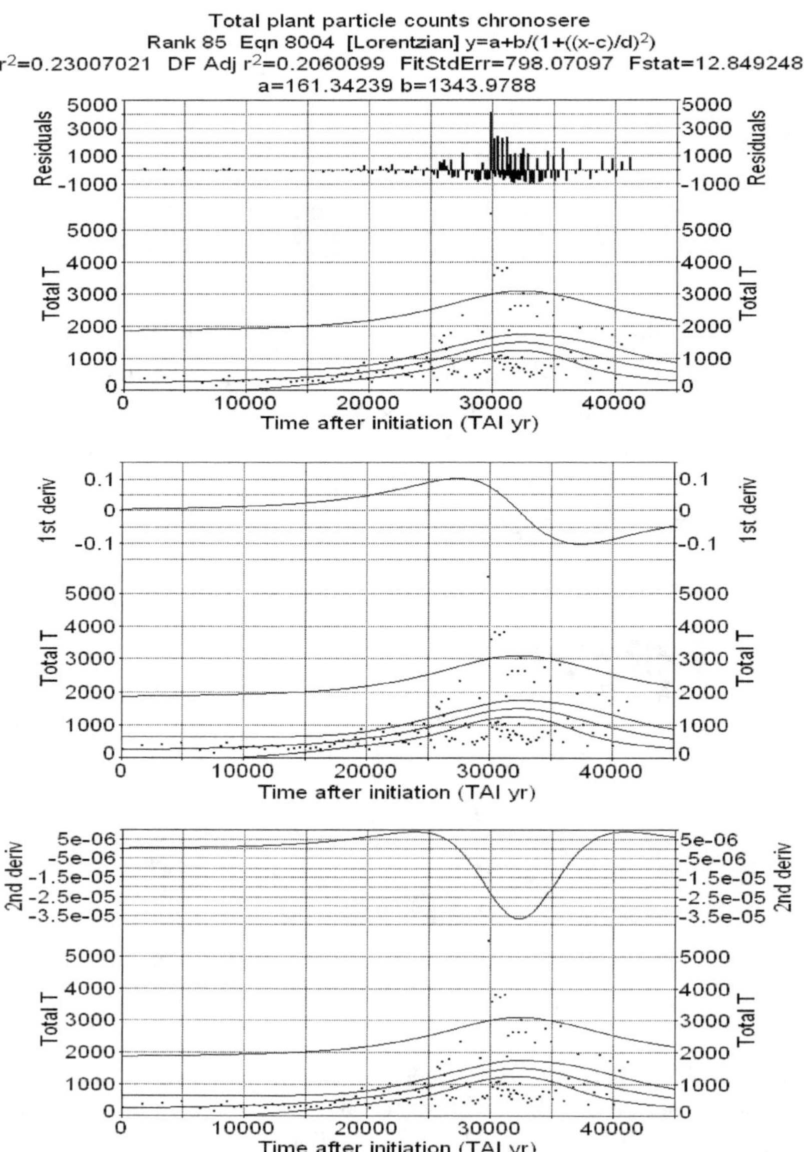

Figure 6B. Residuals are deviations of observed particle counts T from the fitted regression line. This fitted line is a Cauchy-Lorentz function in centre of two 95% confidence belts, one for regression and the other for prediction. Derivatives already defined. Numeric table:

Rank 85 Eqn 8004 [Lorentzian] $y=a+b/(1+((x-c)/d)^2)$

r^2 Coef Det	DF Adj r^2	Fit Std Err	F-value
0.2300702099	0.2060099039	798.07096753	12.849248271

Parm	Value	Std Error	t-value	95% Confidence Limits		P>|t|
a	161.3423878	258.2772091	0.624686895	-349.665396	672.3501717	0.53328
b	1343.978785	255.7215744	5.255633156	838.0273869	1849.930183	0.00000
c	32406.65512	1188.869701	27.25837415	30054.44730	34758.86294	0.00000
d	8549.702809	3049.893990	2.803278684	2515.412852	14583.99277	0.00584

Area Xmin-Xmax	Area Precision		
3.086823e+07	9.53333e-14		
Function min	X-Value	Function max	X-Value
248.80109540	1.733943e-10	1505.3211727	32406.655226
1st Deriv min	X-Value	1st Deriv max	X-Value
-0.102101774	37342.844460	0.1021017744	27470.483557
2nd Deriv min	X-Value	2nd Deriv max	X-Value
-3.67723e-05	32406.623414	9.193066e-06	23856.872989

Procedure	Minimization	Iterations
LevMarqdt	Least Squares	15
r^2 Coef Det	DF Adj r^2	Fit Std Err
0.2300702099	0.2060099039	798.07096753

Source	Sum of Squares	DF	Mean Square	F Statistic	P>F
Regr	24551724	3	8183908.1	12.8492	0.00000
Error	82162328	129	636917.27		
Total	1.0671405e+08	132			

Description: T chronosere

X Variable: Time afterinitiation (TAI yr)

Xmin:	0.0000000000	Xmax:	41138.000000	Xrange:	41138.000000
Xmean:	25563.609023	Xstd:	9209.0461001	Xmedian:	27522.000000
X@Ymin:	7490.0000000	X@Ymax:	29851.000000	X@Yrange:	22361.000000

Y Variable: T

Ymin:	188.00000000	Ymax:	5530.0000000	Yrange:	5342.0000000
Ymean:	994.65413534	Ystd:	899.13279807	Ymedian:	675.00000000
Y@Xmin:	295.00000000	Y@Xmax:	1726.0000000	Y@Xrange:	1431.0000000

Date	Time	File Source
Feb 21, 2014	11:01:57 PM	CLIPBRD.PRN

Total plant particle counts chronosere
Rank 72 Eqn 8004 [Lorentzian] $y=a+b/(1+((x-c)/d)^2)$
r^2=0.24394405 DF Adj r^2=0.2203173 FitStdErr=790.84782 Fstat=13.874097
a=597.23253 b=1135.9965

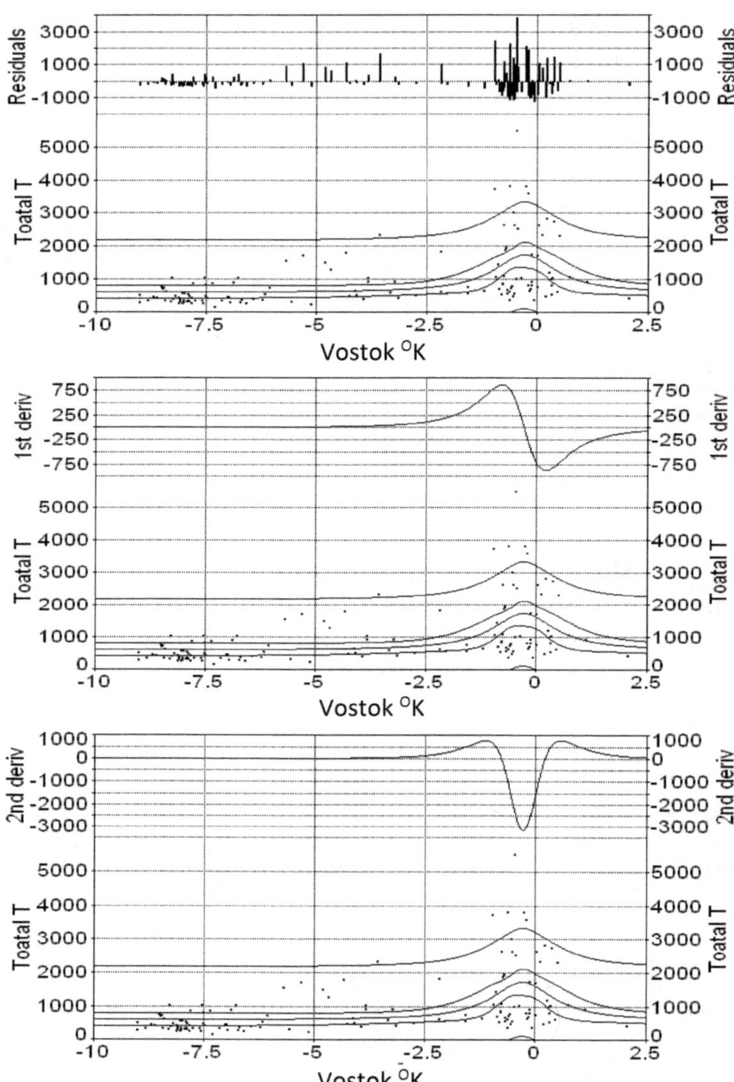

Table 6C. Residuals are deviations of observed particle counts T from the fitted Cauchy-Lorentz function in relation to the Vostok chronosere. Regression line is in centre of two 95% confidence belts, one for regression and the other for prediction. The derivatives follow definitions already given. Numeric table:

Rank 72 Eqn 8004 [Lorentzian] $y=a+b/(1+((x-c)/d)^2)$

r^2 Coef Det	DF Adj r^2	Fit Std Err	F-value
0.2439440461	0.2203172975	790.84781534	13.874097451

Parm	Value	Std Error	t-value	95% Confidence Limits		P>|t|
a	597.2325330	100.1846237	5.961319322	399.0148012	795.4502647	0.00000
b	1135.996498	206.0767888	5.512491264	728.2685240	1543.724472	0.00000
c	-0.29789277	0.150442587	-1.98010932	-0.59554711	-0.00023843	0.04982
d	0.849638731	0.375820589	2.260756212	0.106068491	1.593208970	0.02545

Area Xmin-Xmax	Area Precision		
9222.0199985	1.258538e-11		
Function min	X-Value	Function max	X-Value
607.91085014	-9.019982709	1733.2290308	-0.297892771
1st Deriv min	X-Value	1st Deriv max	X-Value
-868.4295365	0.1926463668	868.42953652	-0.788431908
2nd Deriv min	X-Value	2nd Deriv max	X-Value
-3147.301762	-0.297892543	786.82544134	0.5517461349

Procedure	Minimization	Iterations	
LevMarqdt	Least Squares	16	
r^2 Coef Det	DF Adj r^2	Fit Std Err	r^2 Attainable
0.2439440461	0.2203172975	790.84781534	0.9395003528

Source	Sum of Squares	DF	Mean Square	F Statistic	P>F
Regr	26032258	3	8677419.2	13.8741	0.00000
Error	80681794	129	625440.27		
Total	1.0671405e+08	132			
Lack Fit	74225632	117	634407.11	1.17917	0.39880
Pure Err	6456162.5	12	538013.54		

Description: Total plant particle counts chronosere

X Variable: Vostok OK

Xmin:	-9.020000000	Xmax:	2.0600000000	Xrange:	11.080000000
Xmean:	-3.903458647	Xstd:	3.4969151005	Xmedian:	-3.850000000
X@Ymin:	-7.310000000	X@Ymax:	-0.480000000	X@Yrange:	6.8300000000

Y Variable: T

Ymin:	188.00000000	Ymax:	5530.0000000	Yrange:	5342.0000000
Ymean:	994.65413534	Ystd:	899.13279807	Ymedian:	675.00000000
Y@Xmin:	325.00000000	Y@Xmax:	462.00000000	Y@Xrange:	137.00000000

Date	Time	File Source
Feb 22, 2014	6:58:57 PM	CLIPBRD.PRN

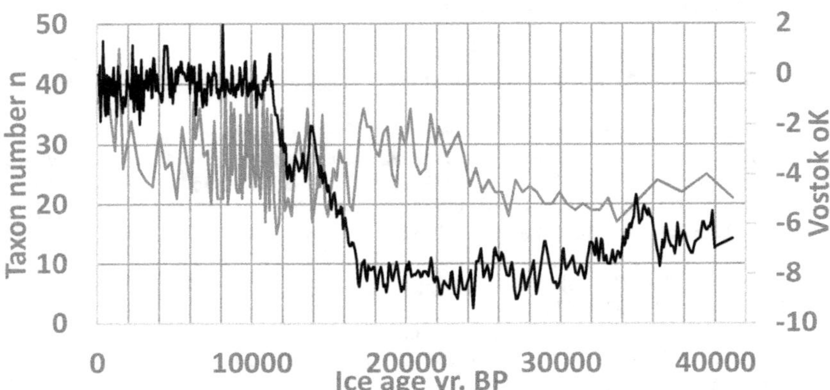

Figure 6D. Graph of the number of palynomorph taxa n within 133 paleo-relevés. The taxon number time scale is Cwynar's carbon age. Vostok graph included for quick reference.

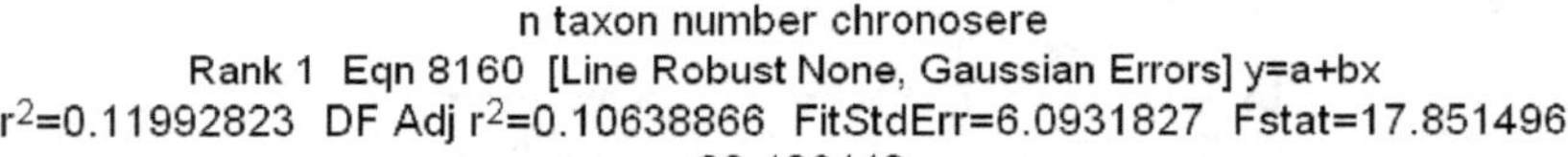

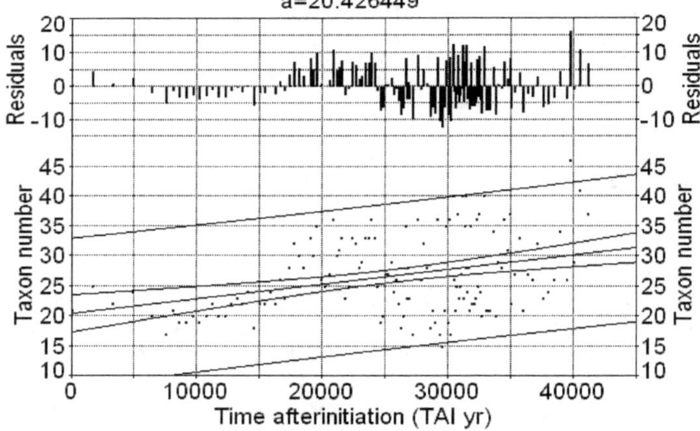

Figure 6E. Residuals are deviations of observed taxon numbers from linear regression. Regression line is embedded within two 95% confidence belts, one for regression and the other for prediction. The regression equation and other regression numerics are in the table:

Rank 1 Eqn 8160 [Line Robust None, Gaussian Errors] y=a+bx

r^2 Coef Det	DF Adj r^2	Fit Std Err	F-value
0.1199282298	0.1063886641	6.0931827068	17.851496478

Parm	Value	Std Error	t-value	95% Confidence Limits		P>|t|
a	20.42644919	1.564129705	13.05930648	17.33222753	23.52067086	0.00000
b	0.000243321	5.75894e-05	4.225103132	0.000129396	0.000357247	0.00004

Area Xmin-Xmax	Area Precision		
1046193.7414	0.0000000000		
Function min	X-Value	Function max	X-Value
20.426449193	1.733943e-10	30.436195632	41138.000000
1st Deriv min	X-Value	1st Deriv max	X-Value
0.0002433212	1582.4101833	0.0002433212	15647.658354
2nd Deriv min	X-Value	2nd Deriv max	X-Value
-2.56262e-16	3397.8194456	2.562618e-16	26962.798596

Procedure	Minimization	Iterations
LevMarqdt	Least Squares	7
r^2 Coef Det	DF Adj r^2	Fit Std Err
0.1199282298	0.1063886641	6.0931827068

Source	Sum of Squares	DF	Mean Square	F Statistic	P>F
Regr	662.77029	1	662.77029	17.8515	0.00004
Error	4863.6207	131	37.126875		
Total	5526.391	132			

Description: n taxon number chronosere

X Variable: Time afterinitiation (TAI yr)

Xmin:	0.0000000000	Xmax:	41138.000000	Xrange:	41138.000000
Xmean:	25563.609023	Xstd:	9209.0461001	Xmedian:	27522.000000
X@Ymin:	29538.000000	X@Ymax:	39720.000000	X@Yrange:	10182.000000

Y Variable: Taxon number

Ymin:	15.000000000	Ymax:	46.000000000	Yrange:	31.000000000
Ymean:	26.646616541	Ystd:	6.4704403493	Ymedian:	25.000000000
Y@Xmin:	21.000000000	Y@Xmax:	37.000000000	Y@Xrange:	16.000000000

Date	Time	File Source
Feb 21, 2014	11:12:41 PM	CLIPBRD.PRN

What can we say about T and n? -

1. The monotonic rising trend is significant in both cases, albeit very slight, arc tan 0.00399637=0.23° for T and arc tan 0.00024251=0.01° for n.

2. The residuals of T are insignificant up to the time when climate change reaches the intense warming phase around year

17000 BP. At that point T starts violent oscillations rising above 1000 particle counts (year 14000 BP) then falling back as the intensive portion of the warming cycle comes to an end.

2. Positive oscillations of n begin in earnest often up to 5 or even 10 units after the bottoming out of the climate cooling cycle. The balancing out of positive and negative deviations after year 17000 BP - whose amplitude reaches high/low points (+10/-10) after year 14000 and maximum (+16) around year 1259 BP - is closely synchronous with temperature oscillations.

3. We deal with two timescales, Vostok age is gas isotope based, Cwynar's is carbon based. It would be foolhardy to expect exact matches of the time steps. For that reason we cannot tell exactly if there is any retro-effect of temperature on residual picks in T and n.

Ghost complexes

Consider 133 nH values, one for each paleorelevé of the Cwynar chronosere, and a single nH value for the 133x89 block which include 3540 nonzero cells and a total count of plant particles 136567. From these we get,

nH for the 133x89 block	Sum of 133 nH values	Difference dnH
16516.01 nats	15287.91 nats	1228.10 nats

There is a surplus of potential energy dnH=1228.10 nats. The surplus signals the existence ghost complexes. What is the value of n in the ghost complex? We find it by iteration:

Iterated n	T	H	Iterated H
94	136567	1228.1	1205.542
95	136567	1228.1	1218.367
96	**136567**	**1228.1**	**1231.193**
97	136567	1228.1	1244.018
98	136567	1228.1	1256.843

The iteration gives n=96. The one resonator potential energy level in the ghost complex is dH= 1228.1/96=12.7927 nats. Corresponding to this P=2.78098E-06, w_{ab}=5.56194E-06 and a stability level better than 99%. The ghost complex materialised is one of 359585 theoretically possible complexes.

The potential energy shadow

A ghost complex comes into analytical view when chronosere elements or segments of elements are linked. I forego the formal symbolism and designate the two partial chronoseres by

A,B and their union by A+B. For example if A is a single paleo-relevé at the beginning of the chronosere then B includes all the others of the chronosere.

My term for the paleorelevé at the locus of which the A and B chronosere segments are joined into A+B is the *pivotal relevé*. The locus of the pivot of the case presented in Table 1 is at 10 k year BP. The chain of paleorelevés from present to 19 k year BP inclusive is my complex A and everything left over after 20 kyr. BP is my complex B. The pivot which may belong to A or to B is excluded. Therefor we can associate the pivot with two dnH quantities. I call these the pivot's energy shadow.

Table 1. Example showing the computation of structural parameters for the potential energy within complexes A, B, A+B and in the ghost complexes G1 and G2. I use the entire Cwynar chronosere (133 paleorelevés and 89 palynomorph taxa) in calculations. Legend: A – the chain of 44 paleorelevés from present to 19 kyr. BP, a block of 95x89=8455 plant particle counts of which 2623 are greater than zero; B – the chain of 89 paleorelevés[25] from 20 kyr. BP to 42 kyr. BP, a block of 37x89=3292 cells of which 2653 contain plant particle count greater than zero; A+B – the entire Cwynar chronosere, extending from present to year 41138 BP including 3540 non-zero plant particle counts. Legend to other terms: n - number of resonators defined as the number of non-zero plant particle counts within the complex A, B or A+B; T - particle count totals within A, B or A+B; nH - energy-based entropy within complex A, B, A+B or G; H=nH/n - energy-based entropy of one resonator in A, B, A+B or G; $w_{ab}=1-w_a-w_b$ – instability parameter (0 to 0.5); P - the probability of a randomly assembled complex with given T and n being exactly the same as the observed A,B or A+B. In the table we give w_{ab} values only for H and dH. The a,b subscript have meaning as the complex actually materialised (a) and at least one other complex with the same T and n.

[25] Numeral match with the number of palynomorphs in pure coincidence.

Pivot left out –

Section	A	B	A+B	G1[1]	G2[1]
n	2623	887	3510	25	33
T	119992	16017	136009	119992	16017
nH or dnH*	12679.47	3477.71	16391.35	234.17*	234.17*
H or dH	4.83396	3.92075	4.66990	9.36669	7.09598
P_a	0.00795	0.01983	0.00937	0.00009[!]	0.00083[!]
$1\text{-}P_a$	0.99205	0.98017	0.99063	0.99991	0.99917
w_a	0.00006	0.00039	0.00009	0.[8]73147	0.[6]68630
w_b	0.98415	0.96074	0.98134	0.99983	0.99834
w_{ab}	0.01578	0.03887	0.01857	0.00017	0.00166
w_{ab} %	3.15668	7.77323	3.71414	0.03421	0.33110

[1] Ghost complex (G) for either A or B.

[*] Values determined by interpolation on the basis of fixed T and fixed dnH[+]

[!] One taxon's potential energy cloud.

[8] Insert 8 zeroes.

[6] Insert 6 zeroes.

The example below shows the iteration of n:

T	Iterated n	Iterated dnH	Error
119992	24	228.4134	-5.72656
119992	**25**	236.9102	2.770217
119992	26	245.367	11.227

We write n=25 for A. We do the following for B:

T	Iterated n	Iterated dnH	Error
16017	32	230.9334	-3.20661
16017	33	237.1356	2.995618
16017	34	243.3076	9.167605

From this we have n=32. We could refine the step size in the iterations and get nominally more accurate results but we have to round to an integer n.

There is a property all should note. I call it homogeneity of proportions T/n. In the case of A, B and A+B the corresponding values are in the following table:

Section	A	B	A+B	dnH nats
n	2623	887	3510	
T	119992	16017	136009	

nH or dnH+	12679.47	3477.71	16391.35	234.17
T/n	47.9	33.3	38.6	

The heterogeneity of the proportions indicates heterogeneity in A+B on the two sides of the pivot. A ghost structure exists and its potential energy level is approximately 234 nats. Should at some pivot nH or dnH be such as in the table below, homogeneity would be the case. No ghost structure could exist:

Section	A	B	A+B	dnH
n	1583	1957	3540	
T	61077	75490	136567	
nH or dnH	7385.743	9130.264	16516.007	0.000
t/n	38.6	38.6	38.6	

A single switch n(A) with n(B) or T(A) with T(B) with T(B) but not both, such as in the next table, will regenerate heterogeneity and a ghost complex would emerge with potential energy level approximately 77 nats:

Section	A	B	A+B	dnH
n	1957	1583	3540	
T	61077	75490	136567	
nH or dnH	8721.517	7717.269	16516.007	77.221
t/n	31.2	47.7	38.6	

The ghost complex's energy level is measurable at any of the 133 paleo relevé as if an energy shadow drifted through time. Figure 7 displays the trace of this drift in relation to the Vostok temperature chronosere.

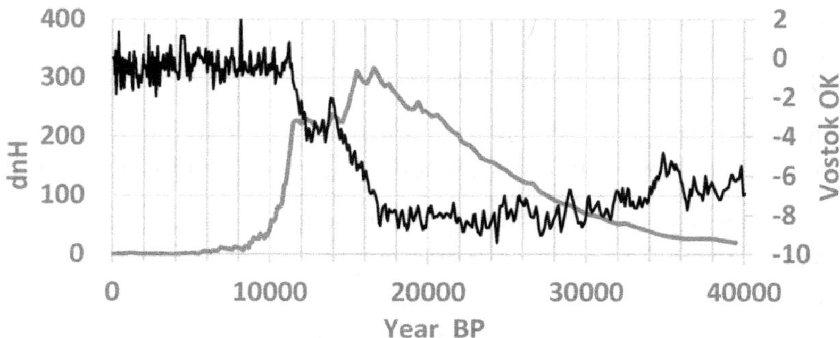

Figure 7A. Z-shaped graph of 637 points is the Vostok temperature chronosere. Superimposed is the dnH graph, the potential energy level of the ghost complex in 133 paleorelevés, each taking its term as pivotal point. The pivot is not included in the calculation of dnH.

The shape function the regression line in Figure 7 is the familiar $f(x) = \dfrac{a + cx + x^2}{1 + bx + dx^2}$. We defined the residuals and gave differential equations earlier in the text.

dnH chronosere

Rank 101 Eqn 7903 y=(a+cx+ex²)/(1+bx+dx²) [NL]

r²=0.91926968 DF Adj r²=0.91609132 FitStdErr=29.616806 Fstat=364.38144

a=10.630447 b=-7.6146734e-05 c=0.00077345216

d=1.5104072e-09 e=-2.8771183e-08

Residuals

dnH

Time after initiation (TAI yr)

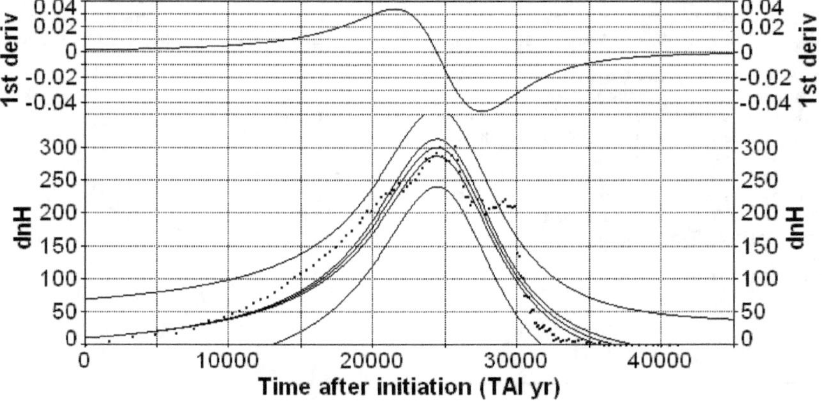

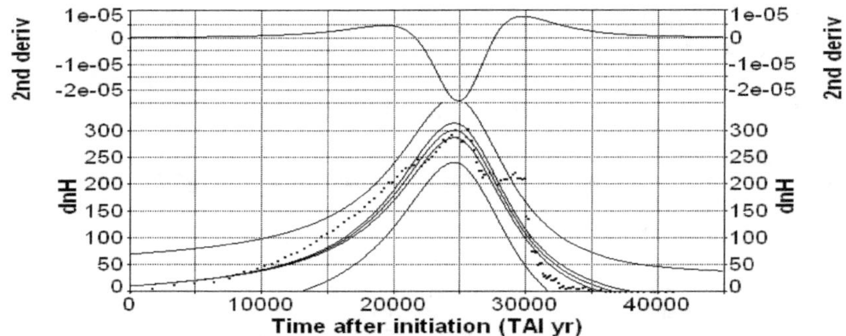

Figure 7B. The energy shadow's trace f(x) for Cwynar's 133 paleorelevés. *Residuals:* deviations of the observed dnH values from the regression line. *Regression line:* embedded in centre within two 95% confidence belts, one for regression (inner) and the other for prediction (outer). Equation and other regression information are in the Table below[26] *1st derivative:* change of f(x) in unit time (velocity). *2nd derivative*: acceleration (above the zero line), deceleration (below the zero line).:

Rank 101 Eqn 7903 $y=(a+cx+ex^2)/(1+bx+dx^2)$ [NL]

r^2 Coef Det	DF Adj r^2	Fit Std Err	F-value
0.9192696812	0.9160913222	29.616805761	364.38143966

Parm	Value	Std Error	t-value	95% Confidence Limits		P>|t|
a	10.63044666	1.61642e-13	6.57655e+13	10.63044666	10.63044666	0.00000
b	-7.6147e-05	4.66248e-07	-163.317955	-7.7069e-05	-7.5224e-05	0.00000
c	0.000773452	1.81491e-09	426166.1429	0.000773449	0.000773456	0.00000
d	1.51041e-09	1.97713e-11	76.39388302	1.47129e-09	1.54953e-09	0.00000
e	-2.8771e-08	6.28333e-10	-45.7896831	-3.0014e-08	-2.7528e-08	0.00000

Area Xmin-Xmax	Area Precision		
3655848.9152	1.575418e-14		
Function min	X-Value	Function max	X-Value
-14.73326362	41137.000000	300.04985789	24504.987185
1st Deriv min	X-Value	1st Deriv max	X-Value
-0.047043134	27597.202204	0.0343882670	21569.712485
2nd Deriv min	X-Value	2nd Deriv max	X-Value
-2.41571e-05	24857.859471	7.848399e-06	29716.153705

Singularities [Data Range]
None
Singularities [All Other]
None

[26] Graphs constructed in the application TableCurve 2D v4. If in doubt how to interpret the output please check out the presentation referenced earlier in the text.

Soln Vector	Covar Matrix	SVD Cond
LvMrq/SVD	SVDecomp	2.589733e+25

Singularities [All Other]
None

Soln Vector	Covar Matrix	SVD Cond
LvMrq/SVD	SVDecomp	2.589733e+25
r^2 Coef Det	DF Adj r^2	Fit Std Err
0.9192696812	0.9160913222	29.616805761

Source	Sum of Squares	DF	Mean Square	F Statistic	P>F
Regr	1278476.3	4	319619.07	364.381	0.00000
Error	112275.86	128	877.15518		
Total	1390752.1	132			

Description: dnH chronosere

X Variable: Time after initiation (TAI yr)

Xmin:	0.0000000000	Xmax:	41137.000000	Xrange:	41137.000000
Xmean:	25563.601504	Xstd:	9209.0332883	Xmedian:	27522.000000
X@Ymin:	0.0000000000	X@Ymax:	25646.000000	X@Yrange:	25646.000000

Y Variable: dnH

Ymin:	0.0000000000	Ymax:	303.38447940	Yrange:	303.38447940
Ymean:	114.94665888	Ystd:	102.64502445	Ymedian:	75.284735170
Y@Xmin:	0.0000000000	Y@Xmax:	1.4942014360	Y@Xrange:	1.4942014360

Date	Time	File Source
Feb 13, 2014	12:57:44 PM	CLIPBRD.PRN

dnH chronosere
Rank 1 Eqn 8160 [Line Robust None, Gaussian Errors] y=a+bx
r^2=0.32948002 DF Adj r^2=0.31908436 FitStdErr=84.288907 Fstat=63.879382
a=50.558756

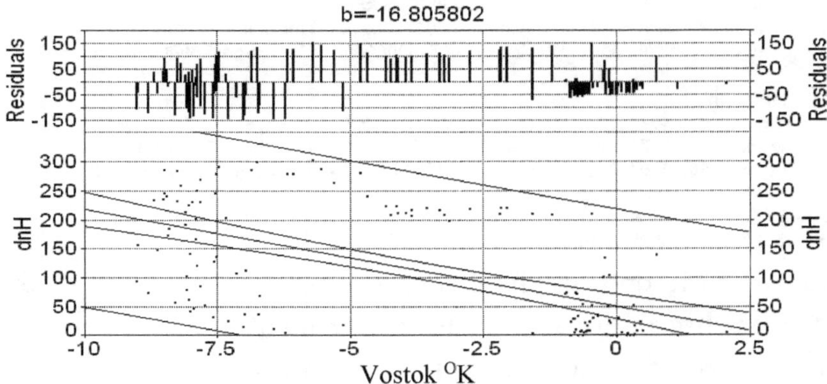

Figure 7C. Regression of dnH on Vostok. See the numeric table below:

Rank 1 Eqn 8160 [Line Robust None, Gaussian Errors] y=a+bx

r^2 Coef Det	DF Adj r^2	Fit Std Err	F-value
0.3294800156	0.3190843569	84.288907137	63.879381697

Parm	Value	Std Error	t-value	95% Confidence Limits		P>|t|
a	50.55875602	10.97682565	4.605954183	28.84241923	72.27509280	0.00001
b	-16.8058018	2.102707607	-7.99245780	-20.9657573	-12.6458463	0.00000

Area Xmin-Xmax	Area Precision		
1208.1958454	0.0000000000		
Function min	X-Value	Function max	X-Value
15.938804272	2.0600000000	202.14679781	-9.019982709
1st Deriv min	X-Value	1st Deriv max	X-Value
-16.80580182	-6.380781644	-16.80580182	-8.068096536
2nd Deriv min	X-Value	2nd Deriv max	X-Value
-5.65211e-08	-6.804000000	5.652106e-08	-8.075436578

Procedure	Minimization	Iterations	
LevMarqdt	Least Squares	6	
r^2 Coef Det	DF Adj r^2	Fit Std Err	r^2 Attainable
0.3294800156	0.3190843569	84.288907137	0.9531665952

Source	Sum of Squares	DF	Mean Square	F Statistic	P>F
Regr	453838.72	1	453838.72	63.8794	0.00000
Error	923600.58	130	7104.6199		
Total	1377439.3	131			
Lack Fit	859090.41	118	7280.4272	1.35428	0.28858
Pure Err	64510.173	12	5375.8477		

Description: dnH chronosere

X Variable: Vostok oK

Xmin:	-9.020000000	Xmax:	2.0600000000	Xrange:	11.080000000
Xmean:	-3.883106061	Xstd:	3.5023205246	Xmedian:	-3.710000000
X@Ymin:	-0.020000000	X@Ymax:	-5.710000000	X@Yrange:	5.6900000000

Y Variable: dnH

Ymin:	0.0040818840	Ymax:	303.38447940	Yrange:	303.38039752
Ymean:	115.81746691	Ystd:	102.54171770	Ymedian:	77.708014440
Y@Xmin:	95.053046330	Y@Xmax:	6.1230403660	Y@Xrange:	88.930005964

Date	Time	File Source
Feb 15, 2014	12:29:39 PM	CLIPBRD.PRN

Consider the dnH graphs in Figure 7B. We may think of the Cwynar chronosere as the path of the pivot moving with time out of the past in the direction of present. At each pivotal point where the A,B chains link up there is a new Ghost complex whose potential energy state is dnH. The setup is such that the A and B both join to the pivot. So the ghost potential energy cloud associated with the pivot is a single dnH quantity. The dH value can be different (see Table 1).

There are separate graphs for the 1st and 2nd derivatives in Figure 7B. Whenever the graph crosses the zero line upward the rate of change in dnH or the acceleration of change is positive. Whenever the graph crosses downward the opposite happens. Regarding interpretation of the extreme points consider the following:

Extreme points of $f(x) = \frac{a + cx + ex^2}{1 + bx + dx^2}$	Minimum or sign change	TAI yr.	Maximum or sign change	TAI yr.
dnH nat	-15	41137	300	24505
1st deriv	-0.047	27597	0.034	21569
	- to +	14 kyr	+ to -	24 yr
2nd deriv	-0.000024	24857	0.0000078	29716
	+ to -	22 kyr	- to +	28 kyr

What do we deduce from the graphs? --
1. dnH parameter is on a rising trend and attains maximum at 24505 yr. AI (16633 yr. BP) just as the climate warming cycle gets into high gear.
2. 1st derivative reaches minimum in year 27597 AI (13541.yr BP). This is the point where the Vostok regression graph has inflexion point and the warming rate starts slowing down.
3. Sign of the 2nd derivative changes from plus to minus at about 22 kyr. AI and from minus to plus at about 28 kyr. AI brackets the maximum of dnH.

What else do the graph tell us?

1. If we were interested in building a potential energy model for Beringia with comparable objectives as the energy model of Huang Xi and Zu Yuangang (2001) and Huang Xi (2003) the regression generating functions for temperature and dnH could be among its components.

2. Ghost complexes in the Cwynar chronosere thrive where the great climate cooling cycle gives way for the climate warming cycle. This is in year 17 k BP at which the Cwynar chronosere is fractured into two most distinct partial chains. The Vostok chronosere's has 2nd derivative around 17 kyr. BP and it bottoms out close to 22ky BP. I call the 17kyr. BP point a quantum point at which energy dynamics undergoes fundamental change.

More on dnH

I used in earlier papers the acceleration/deceleration metric based on finite intervals in the form $A=\frac{D^2y}{Dt^2}=\frac{Dy_1/t_1-Dy_2/t_2}{t_1+t_2}$. Dy/t, in which y=dnH, indicates velocity, the rate of change in the value of dnH over time t. The A expression is acceleration, the rate of change in velocity. Its oscillogram is in Figure 8. Clearly, the oscillation of A goes through periods of apparent stability and explosive change. The large scale version of the A graph in Figure 8 indicates millennial periods of stability. Violent oscillations follow sustained climate warming.

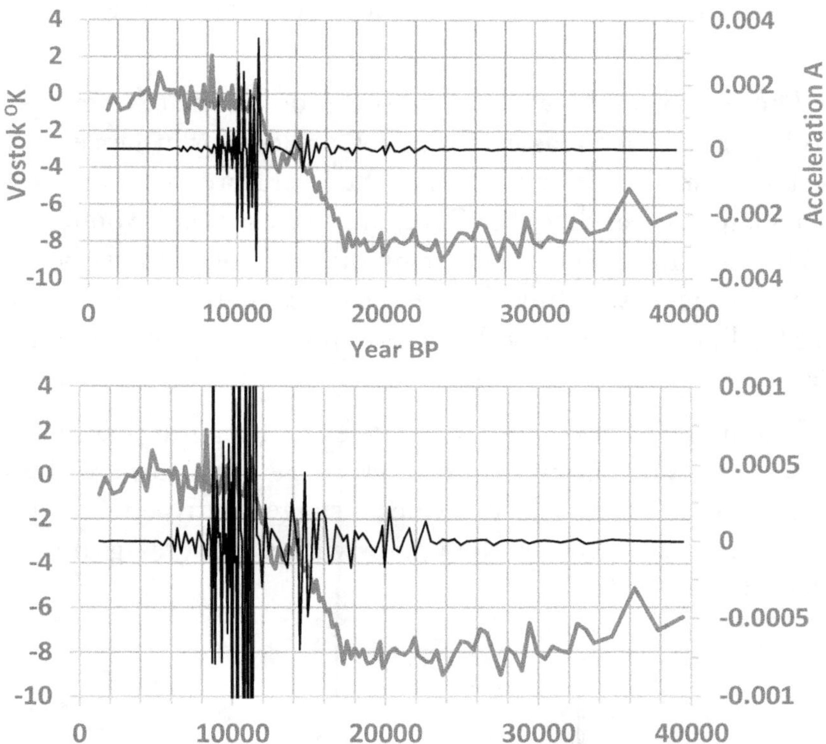

Figure 8. Graph of the acceleration (A) in two magnifications. The number of points is 133. Each point represents the dnH value of one paleorelevé. Oscillations upward indicate positive acceleration. Oscillations downward indicate negative acceleration or deceleration. Acceleration refers to the rate of change in the velocity of dnH transitions per year. The Vostok graph of temperature differences (here only 133 matching points) is included for quick reference.

What is the utility of the A based oscillogram on dnH? It is the direct portrayal of the sensitivity of the ghost complex's potential energy state to conditions which change with changes in the global tropospheric temperature. In this regard first intuitions suggest that the A signal is a delayed energy state response to temperature peaks. This is the obvious yet a specious deduction for several reasons:

1. The Vostok and Cwynar time scales do not match exactly. I already mentioned reasons for this. Consider for instance the A peak in year 8726 BP and then the temperature peak four centuries later in year 8295 BP. They are so close that I would have to know the measurement errors in the two isotope dates before I could suggest which of the two peaks came first in real time. But I believe that the A peak is a retro response to a temperature peak.

2. Claiming common sense in my defence, I suggest that in universal time Vostok temperature peaks are precursors to the A peaks. I feel I am not far out when I suggest retro effects from zero to centuries. I drew the graphs with a 500 yr. lag in retro effects in Figure 9.

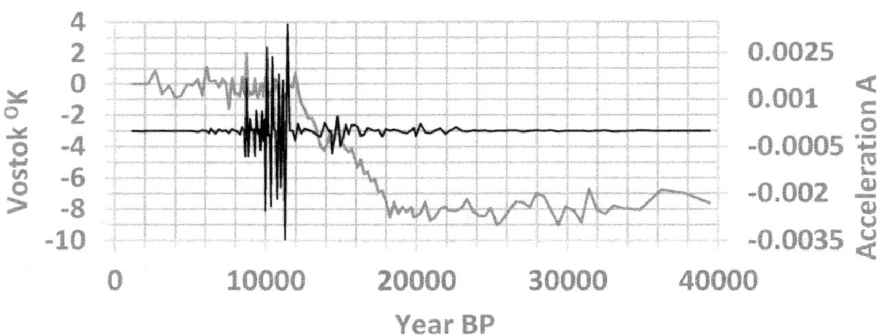

Figure 9. Modified version of Figure 8. The zero point of the A graph is set to 500 yr. BP on the Cwynar time scale. In this position the major peaks are lined up on the Vostok and A graphs in a common sense order, A peaks follow the temperature peaks.

The temperature maximum 2.06 °K at yr. 8726 BP rose from the low of -0.87 °K in 91 years at a 100-year rate of 3.2 °K. The temperature dropped to -0.64 °K by year 8043 BP in 92 years at the 100 year rate of 2.9 ^{O}K. The temperature rise at the rate of 3.2 ^{O}K in 100 years must have devastated the Beringian vegetation and probably melted deep the permafrost in the summer seasons.

Others may decide if my conjecture of a 500-year retro effect is about right, too much or too little. I only say with confidence that making A of dnH, an emergent vegetation property, precursor of conditions that lead to concomitant climate warming would be an oxymoron.

Multiscale correlation r(dnH,^{O}K)

It is an obvious next step to look for direct numerical relationship of dnH and the Vostok temperature differences chronosere. In my 2012 monograph *Statistical multiscaling in dynamic ecology* I used *multiscale correlation analysis*. I will use this method again but I do not apply at this time equal time step transformations prior to correlation analysis. I give an example for equal step transformation on p. 269 in *Statistical ecology*.[27]

Multiscale correlation analysis goes like this:

1. Take two chronoseres with matching time steps:[28]

Series dnH: 2 1 4 6 5   Series ^{O}K: 1 0 -3 -2 1
Note "present" is left and past is right. This is the *zero lag* setup. Matched pairs are 5 and 1, 6 and -2, 4 and -3, 1 and 0, 2 and 1.

2. Create a new series one step shorter than the previous one such that the second last dnH value matches the last ^{O}K value. Call this the *lag 1* setup. In this the matching pairs are: 6 and 1,

[27] 2010 rev. 2014 Online Edition: https://createspace.com/3476529

[28] If the time steps do not match in the two chronoseres my equal time step transformation should be used.

4 and -2, 1 and -3, 2 and 0. If we had long series such as the Cwynar chronosere loosing points at the ends should not make much difference.

3. Consider the Cwynar chronosere, do what has been done so far and go on creating lag 2, lag 3 and other higher lag setups in the same manner by moving the ^{O}K series to the left step by step while holding the dnH series fixed in one position.

4. What has been done so far qualifies as the *unit block* setup. Go on creating the *block 2* setup by using a sliding window of two time points, averaging the corresponding dnH and ^{O}K duplets within each window and replacing the original elements by their average. For example, the series is 2, 1, 4, 6, 5 at unit block size becomes 1.5, 2.5, 5, 5.5 at block size 2. Continue creating new series of block 3, 4, etc. with lag 1, 2, etc.

5. At this point if we had, say 11 values for lag and 20 block sizes there is a total of 220 n-valued chronosere duplets.

6. Now we are ready at a point to calculate the product moment correlation coefficient, symbolically r(dnH,^{O}K), for each of the chronosere duplets. This would give us 220 values for r(dnH,^{O}K) – one for each case of lag for each block size. We may graph correlations by lag over block size. But we are not yet finished.

At this point in the process the r(dnH,^{O}K) graph for a given lag provides information about the retro effect at the scale of lag and time step size at the block size.

7. There is one more very tedious task to complete if we want to do the exercise for real. This task is random resampling of the chronosere duplets for each lag a very large number of times with randomly chosen block sizes, minimum 5 consecutive time points long, never longer than a given maximum. The blocks

are randomly placed on the chronosere duplets. Each time the entire analysis is repeated. The end-products are expected frequencies for r(dnH,°K) by sign (F+ or F-) for given lags at each bock size.[29]

I am presenting the numerical results in Figure 10 and key results in Table 2 for lag 0 and 6.

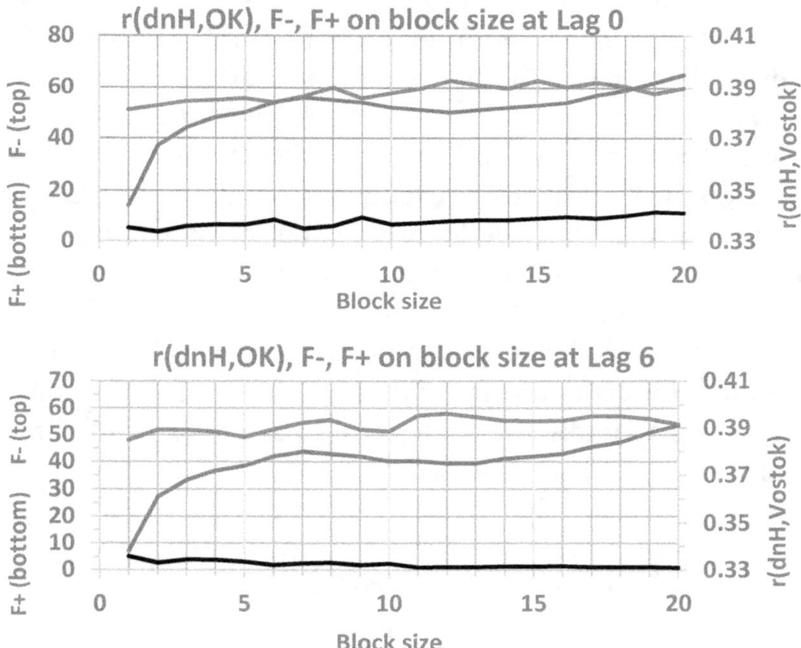

Figure 10. Results of multiscale correlation analysis presented for lag 0 and lag 6. Legend: r(dnH,°K) – the observed product moment correlation coefficient at different block sizes laid on the dnH and Vostok temperature differences coenosere duplet at lag 0 or 6; F- - expected frequency of negative

[29] I created source code in TrueBasic for multiscale correlation analysis some years ago as visiting professor of UFRGS in Prof. Dr. Valério De Patta Pillar's lab in Porto Alegre, Rio Grande do Sul, Brazil.

correlation under assumption of the rule of chance, F+ - expected frequency of positive correlations. Equal time step width adjustment not applied.

The dominance of F- over the total length of the chronosere duplet is surprising. This reminds me of results in earlier paper (Orlóci et al. 2006, Orlóci 2009 and references therein) in which we applied multiscale correlation analysis to the Vostok temperature differences chronosere and Cwynar plant particle based velocity chronosere among a large number of others from climatically different biomes, and found that in arid climates the F- graph is dominant in contrast with humid climates where the F+ graph dominates. The Hanging Lake site showed F+ dominance on the basis of velocity following equal time step adjusted plant particle counts (see Figure11). This is unlike the present dnH based comparison in which F- is dominant. This could have been predicted from the Vostok and dnH graphs in Figure 7A just by observing the upward trend of dnH in the climate cooling cycle and the sharp downward trend in the climate warming cycle. This may imply the value of dnH as indicator of climatic aridity.

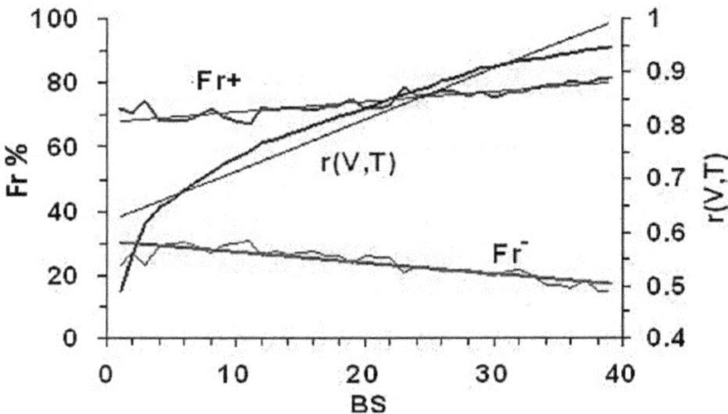

Figure 11. Multiscale correlation analysis of Vostok temperature differences (T in graph) and the Cwynar velocity chronosere (V). Equal time step adjustment was applied. Graph adapted from Orlóci et al. (2006).

Newtonian force and dnH

Force is mass times acceleration. In our particular case it is F=MA. In this M=T/n, the plant particle density of one taxon in the paleorelevé; $A = \frac{D^2y}{Dt^2}$, the second derivative of y written for finite differences; and y=f(x), the value of dnH at time point x on the TBP axis. What does F really represents in this case? Think of F as the level of inertia of the ghost complex's parameter dnH to resist change (A negative) or its readiness (A positive) to change. Graphs are in Figure 12.

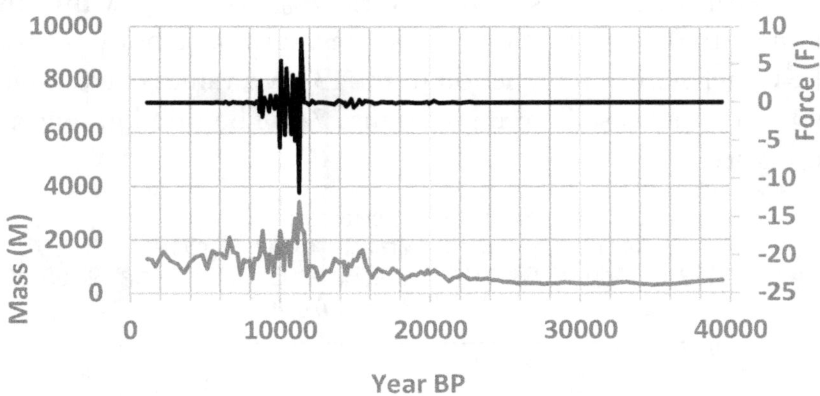

Figure 12. The potential force associated with the dnH level of Cwynar's 133 paleorelevés.

Potential energy and work

Continuing with the ghost complex's potential energy dnH as an example and the force equation F=MA. Clearly, if dnH represents force F then it also represents work W=Fs assuming that F does not change. In this s is the displacement over which we want to define W. But F does change and for that reason we should write $W = \int_{x_1}^{x_2} f(x)_F f(x)_{dnH} dx$. In this x_1 and x_2 are two time points, $f(x)_{dnH}$ is the Cauchy-Lorentz function, and s is distance on the curve between x_1 and x_2. W is the work needed to move the ghost cloud from the pivotal paleorelevé at time point x_1 to the pivotal paleorelevé at time point x_2.

The diligent student will no doubt try to expand this and the previous sections and find actual numerical values.

dnH-based instability index

We recall that the instability metric w_{ab} has values within the limits 0 to 0.5 and that 0.5-w_{ab} is a measure of stability. I fitted the Cauchy-Lorentz function to the 131 w_{ab} values corresponding to the dnH oscillogram in Figure 7A. The fitted graphs are in Figure 13A.

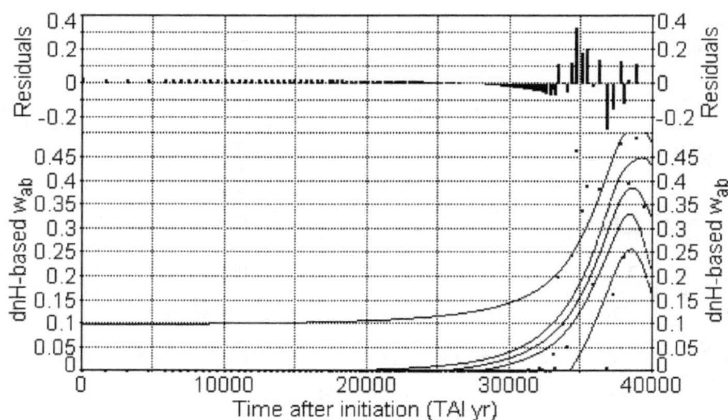

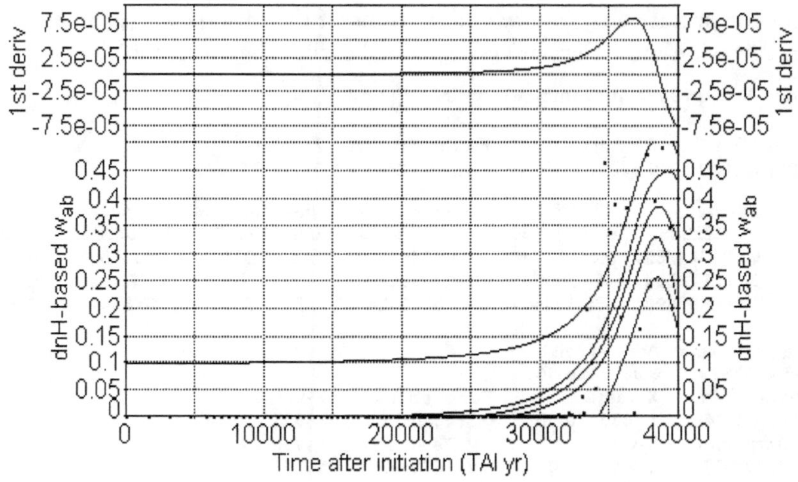

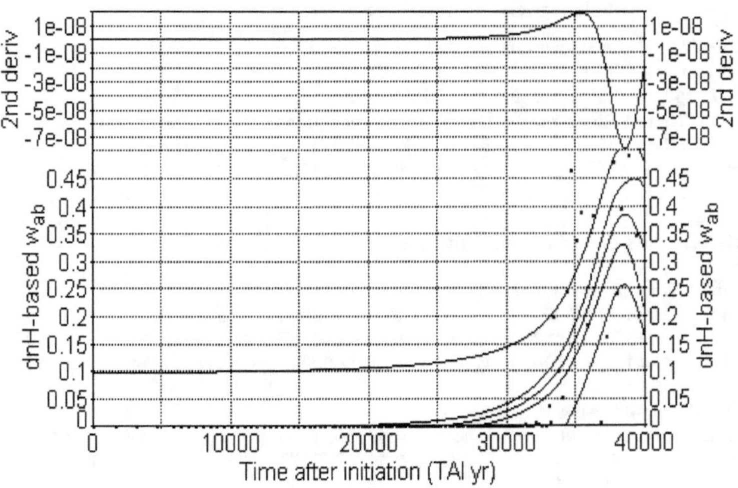

Figure 13A. Regression analysis of the dnH-based w_{ab} values corresponding to Figure 7A. The number of points is 131. Each in turn served as pivot. The pivot is not considered in the computation of the w_{ab} value. *Regression line*: middle line within two 95% confidence belts, one for regression and the other for prediction. *Residuals:* deviations of the observed w_{ab} values from the regression line. *Derivatives:* explained earlier. Numeric table:

Rank 190 Eqn 8004 [Lorentzian] $y=a+b/(1+((x-c)/d)^2)$

r^2 Coef Det	DF Adj r^2	Fit Std Err	F-value
0.6988179011	0.6893318507	0.0585872594	98.997352615

Parm	Value	Std Error	t-value	95% Confidence Limits		P>|t|
a	-0.02220457	0.008440460	-2.63073034	-0.03890546	-0.00550368	0.0095
b	0.407044752	0.028728468	14.16868987	0.350200570	0.463888934	0.0000
c	38615.28581	454.7895414	84.90803393	37715.40700	39515.16462	0.0000
d	3214.905898	475.8945016	6.755501245	2273.267320	4156.544477	0.0000

Area Xmin-Xmax	Area Precision		
1405.4725623	3.382332e-10		
Function min	X-Value	Function max	X-Value
-0.019402624	1.663047e-10	0.3848401786	38615.285785
1st Deriv min	X-Value	1st Deriv max	X-Value
1.441223e-07	2.001875e-09	8.223672e-05	36759.169520
2nd Deriv min	X-Value	2nd Deriv max	X-Value
-7.87654e-08	38615.284436	1.969136e-08	35400.382285

Procedure	Minimization	Iterations
LevMarqdt	Least Squares	33
r^2 Coef Det	DF Adj r^2	Fit Std Err
0.6988179011	0.6893318507	0.0585872594

Source	Sum of Squares	DF	Mean Square	F Statistic	P>F
Regr	1.0194154	3	0.33980514	98.9974	0.0000
Error	0.43935577	128	0.003432467		
Total	1.4587712	131			

Description: dnH-based wab chronosere

X Variable: Time after initiayion (TAI yr)

Xmin:	0.0000000000	Xmax:	39456.000000	Xrange:	39456.000000
Xmean:	24076.265152	Xstd:	8968.1294737	Xmedian:	25982.000000
X@Ymin:	24160.000000	X@Ymax:	38870.000000	X@Yrange:	14710.000000

Y Variable: dnH-based wab

Ymin:	0.0000000000	Ymax:	0.4935036850	Yrange:	0.4935036850
Ymean:	0.0345288149	Ystd:	0.1055256273	Ymedian:	0.0000000000
Y@Xmin:	0.0026653140	Y@Xmax:	0.3481198730	Y@Xrange:	0.3454545590

Date	Time	File Source
Mar 11, 2014	4:27:09 PM	CLIPBRD.PRN

The graphs of Figure 13A can be viewed as traces of a single ghost complex moving through time from left to right while its instability state undergoes continuous change. It is clear that instability is low-level and stable during the climate cooling cycle. It becomes explosive as climate warming progresses. The residuals indicate this. Instability attains maximum in year 38615 AI (year 2523 BP). The maximum of the 1st derivative and its minimum brackets the maximum instability point; the 2nd derivative's minimum points directly down onto it. The w_{ab} graphs add further support to the conjecture that climate warming is a strong destabiliser of vital vegetation parameters.

The Cauchy-Lorentz function's parameters suggests a good fit to the w_{ab} chronosere. But the smoothness imparted is for the dominant trend. Immense turbulence occur in w_{ab}'s oscillations in tandem with similar oscillations of tropospheric temperature. The residuals indicate this for f(x) and Figure 13B for the finite deviation 2nd derivative[30].

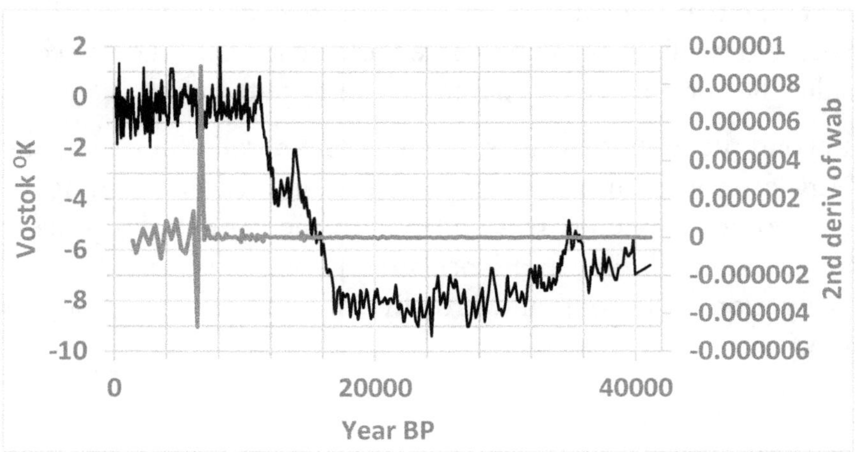

---

30 $A=\frac{\Delta^2 w_{ab}}{\Delta t^2}=\frac{Dw_{ab1}/t_1-Dw_{ab2}/t_2}{t_1+t_2}$

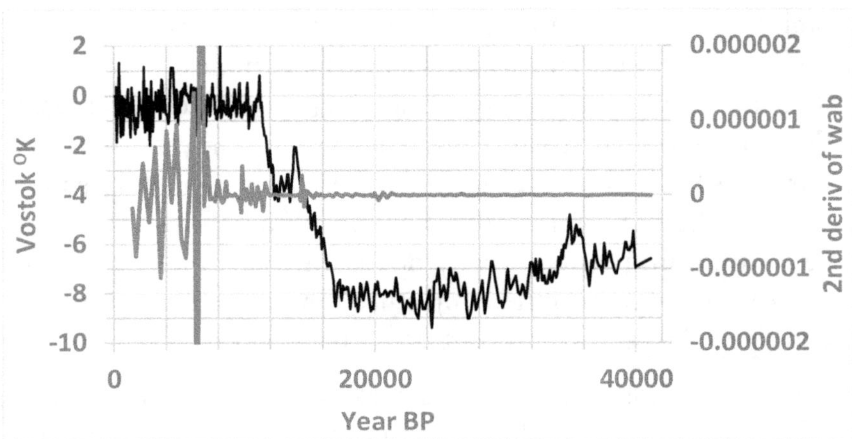

Figure 13B. Graphs of the finite deviation expression of the w_{ab}'s 2nd derivative. Two levels of enlargement are shown. The Vostok temperature differences is included for quick reference. Note the 1.6k year lag of the apparent retro-effect at its sharpest.

Is it true that the whole is greater than the sum of its parts? Of course! I would feel having wasted information if I did not consider the ghost complex's potential energy at the pivot.

Beyond the examined cases, the idea of emergent potential energy cannot be limited to a single unidirectional joining of chronosere segments. It is endemic to the community process on any scales in time and space as short and local as Kernerian facilitation or as long and broadly based as the phylogeny of species.

nH oscillations

Having completed the discussion of dnH, the potential energy state of ghost complexes, now we turn to the nH, the potential energy state of the paleorelevé itself. Oscillogram and fitted curves are placed into Figure 14.

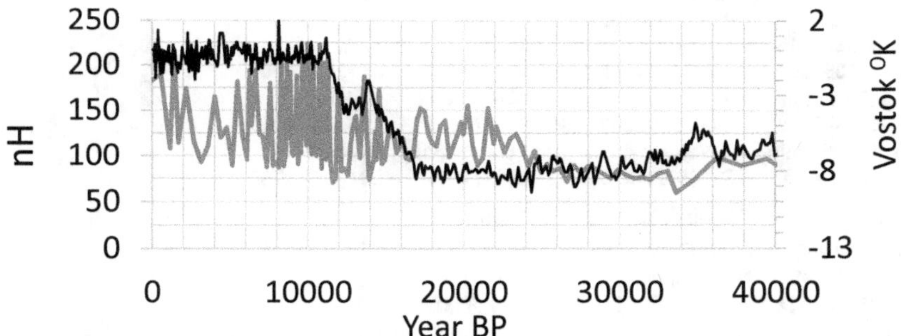

Note change of time scale below to TAI.=41138-TBP yr.

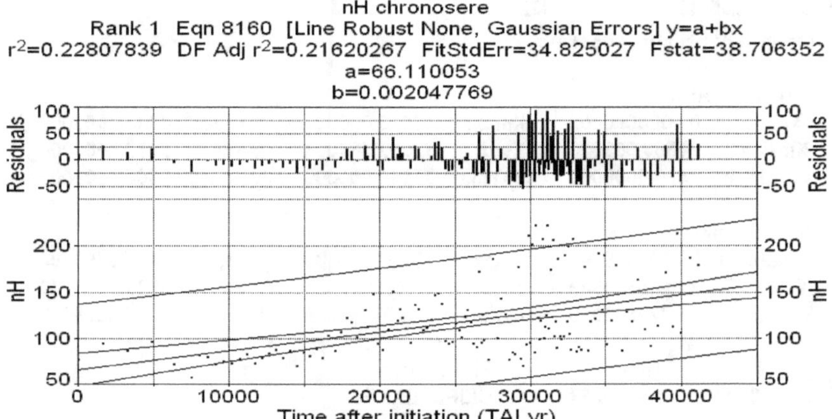

Figure 14A. *Top diagram.* Observed nH (not dnH) of Cwynar's plant particle chronosere and Vostok temperature oscillogram juxtaposed on their time scales. *Bottom graph: deviations above,* straight regression line below fitted to nH embedded within two inclusive 95% confidence belts, inner for regression, outer for prediction of nH. Numerical table:

Rank 1 Eqn 8160 [Line Robust None, Gaussian Errors] y=a+bx

r^2 Coef Det	DF Adj r^2	Fit Std Err	F-value
0.2280783914	0.2162026744	34.825026787	38.706351716

Parm	Value	Std Error	t-value	95% Confidence Limits		P>|t|
a	66.11005320	8.939639839	7.395158462	48.42531314	83.79479326	0.00000
b	0.002047769	0.000329147	6.221442897	0.001396638	0.002698900	0.00000

Area Xmin-Xmax	Area Precision		
4452390.9956	0.0000000000		
Function min	X-Value	Function max	X-Value
66.110053201	1.733943e-10	150.35117465	41138.000000
1st Deriv min	X-Value	1st Deriv max	X-Value
0.0020477690	8485.2760293	0.0020477690	110.54428363
2nd Deriv min	X-Value	2nd Deriv max	X-Value
-2.05009e-15	39566.604070	1.025047e-15	28443.314766

Procedure	Minimization	Iterations
LevMarqdt	Least Squares	6
r^2 Coef Det	DF Adj r^2	Fit Std Err
0.2280783914	0.2162026744	34.825026787

Source	Sum of Squares	DF	Mean Square	F Statistic	P>F
Regr	46942.386	1	46942.386	38.7064	0.00000
Error	158874.51	131	1212.7825		
Total	205816.89	132			

Description: nH chronosere

X Variable: TAI yr

Xmin:	0.0000000000	Xmax:	41138.000000	Xrange:	41138.000000
Xmean:	25563.609023	Xstd:	9209.0461001	Xmedian:	27522.000000
X@Ymin:	7490.0000000	X@Ymax:	31140.000000	X@Yrange:	23650.000000

Y Variable: nH

Ymin:	58.601329860	Ymax:	222.83332180	Yrange:	164.23199194
Ymean:	118.45841949	Ystd:	39.486945668	Ymedian:	109.03233940
Y@Xmin:	77.221838180	Y@Xmax:	179.57160430	Y@Xrange:	102.34976612

Date	Time	File Source
Feb 14, 2014	5:10:18 PM	CLIPBRD.PRN

The steady rise of nH form initiation 41138 years ago to the left is noted and so is the fact that the rise is statistically significant. Note the violent oscillations within the last phase of the climate warming cycle. The regression is low precision, but the F-ratio is highly significant. The actual slope of the regression line is a mere arc tan 0.00203617 = 0.12°. The predicted linear rise of nH over the entire length of the chronosere is 0.2 nats per 100 years.

Figure 14B gives a direct view of the nH scatter over 133 point on the Vostok temperature scale. The real slope of the regression line in this case is arc tan 4.17845432 = 76.5°. The predicted rise in nH per one °K is 4.3 nats.

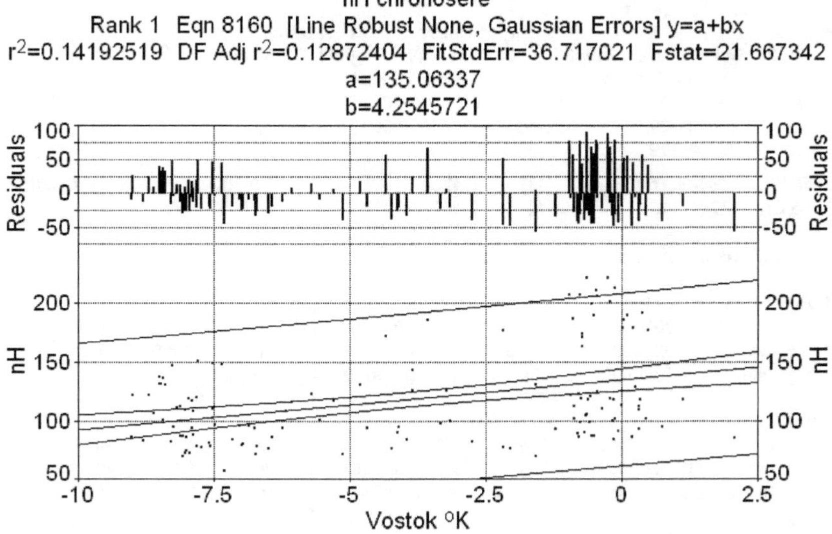

Figure 14B. Linear regression of the paleorelevés' nH on the Vostok temperature differences scale and residuals. Regression line is bound by two 95% confidence belts, one for regression and the other prediction of nH. The regression line is rather inefficient, yet still statistically significant. Numeric table:

Rank 1 Eqn 8160 [Line Robust None, Gaussian Errors] y=a+bx

r^2 Coef Det	DF Adj r^2	Fit Std Err	F-value
0.1419251944	0.1287240435	36.717020777	21.667342215

Rank 1 Eqn 8160 [Line Robust None, Gaussian Errors] y=a+bx

r^2 Coef Det	DF Adj r^2	Fit Std Err	F-value
0.1419251944	0.1287240435	36.717020777	21.667342215

Parm	Value	Std Error	t-value	95% Confidence Limits		P>|t|
a	135.0633744	4.781395067	28.24769184	125.6046344	144.5221144	0.00000
b	4.254572054	0.914014450	4.654819246	2.446433443	6.062710665	0.00001

Area Xmin-Xmax	Area Precision		
1332.4526976	0.0000000000		
Function min	X-Value	Function max	X-Value
96.687208069	-9.019982709	143.82779286	2.0600000000
1st Deriv min	X-Value	1st Deriv max	X-Value
4.2545720537	2.0581868909	4.2545720537	-7.886073779
2nd Deriv min	X-Value	2nd Deriv max	X-Value
-2.82606e-08	-7.913953221	2.826055e-08	0.8765517351

Procedure	Minimization	Iterations	
LevMarqdt	Least Squares	7	
r^2 Coef Det	DF Adj r^2	Fit Std Err	r^2 Attainable
0.1419251944	0.1287240435	36.717020777	0.9137188511

Source	Sum of Squares	DF	Mean Square	F Statistic	P>F
Regr	29210.602	1	29210.602	21.6673	0.00001
Error	176606.29	131	1348.1396		
Total	205816.89	132			
Lack Fit	158848.17	119	1334.8586	0.902027	0.64042
Pure Err	17758.118	12	1479.8432		

Description: nH chronosere

X Variable: Vostok oK

Xmin:	-9.020000000	Xmax:	2.0600000000	Xrange:	11.080000000
Xmean:	-3.902849624	Xstd:	3.4964506906	Xmedian:	-3.850000000
X@Ymin:	-7.310000000	X@Ymax:	-0.650000000	X@Yrange:	6.6600000000

Y Variable: nH

Ymin:	58.601329860	Ymax:	222.83332180	Yrange:	164.23199194
Ymean:	118.45841949	Ystd:	39.486945668	Ymedian:	109.03233940
Y@Xmin:	87.403624660	Y@Xmax:	86.382092840	Y@Xrange:	1.0215318200

Date	Time	File Source
Feb 14, 2014	5:18:16 PM	CLIPBRD.PRN

What else can we tell concerning the 133 point nH chronosere? Three points:

1. nH oscillations intensify after the cooling cycle reaches its nadir. The oscillations become explosive when the slow warming process builds up a critical retro effect. Dampened nH oscillations in the vicinity of extreme temperature peaks suggest dramatic compositional transitions brought on the vegetation by sudden climate warming.

2. A positive relationship of nH and temperature rise suggests that climate warming is the ultimate driver of nH's rising trend.

3. Characterization of the nH and climate change relationship in more precise time-point to time-point terms remains elusive for reasons of discrepant scales which I already discussed.

H oscillations

The energy level of one taxon in the relevé, H=nH/n, is portrayed by oscillogram and fitted regression graphs in Figure 15. Each graph spans the entire Cwynar chronosere of 133 paleo-relevés.

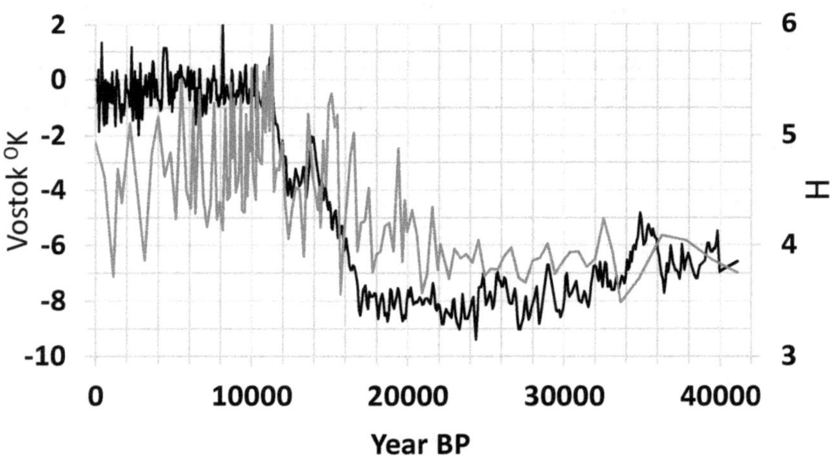

Figure 15A. Graph of the observed one taxon potential energy H in 133 paleorelevés juxtaposed with the graphs of the 637 valued Vostok temperature differences chronosere.

H cronosere

Rank 110 Eqn 8004 [Lorentzian] $y=a+b/(1+((x-c)/d)^2)$

r^2=0.45610286 DF Adj r^2=0.43910607 FitStdErr=0.42386666 Fstat=36.059066

a=3.5696582 b=1.2484699

c=32000.874 d=9574.8109

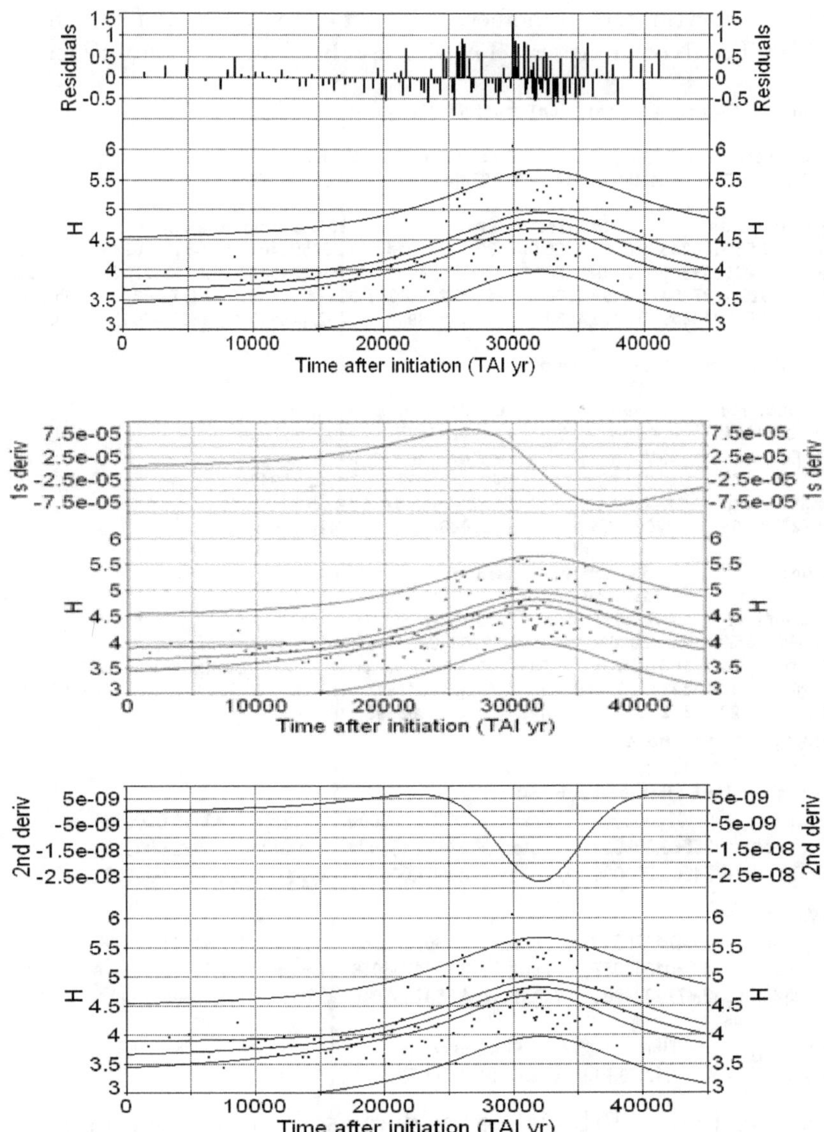

Figure 15B. The Cauchy-Lorentz function fitted to 133 H values. Time scale is TAI=41138-TBP. *Residuals:* deviations of the observed H from the regression line. *Regression line:* centre line within two 95% confidence belts, one for regression (inner) and another for prediction (outer). *1st and 2nd residuals:* as identified. Note, time scale TAI=41138-TBP year. Numerical table:

Rank 110 Eqn 8004 [Lorentzian] y=a+b/(1+((x-c)/d)2)

r2 Coef Det	DF Adj r2	Fit Std Err	F-value
0.4561028571	0.4391060714	0.4238666631	36.059065784

Parm	Value	Std Error	t-value	95% Confidence Limits		P>|t|
a	3.569658248	0.157723014	22.63245014	3.257599402	3.881717094	0.00000
b	1.248469859	0.153017020	8.159026082	0.945721937	1.551217782	0.00000
c	32000.87394	746.4197663	42.87248997	30524.06415	33477.68373	0.00000
d	9574.810859	2063.488388	4.640108913	5492.148545	13657.47317	0.00001

Area Xmin-Xmax	Area Precision		
171259.35835	7.23908e-09		
Function min	X-Value	Function max	X-Value
3.6722419142	1.733943e-10	4.8181281075	32000.877318
1st Deriv min	X-Value	1st Deriv max	X-Value
-8.46915e-05	37528.891103	8.469149e-05	26472.859108
2nd Deriv min	X-Value	2nd Deriv max	X-Value
-2.72363e-08	32000.979321	6.809068e-09	22426.070666

Procedure	Minimization	Iterations
LevMarqdt	Least Squares	12
r2 Coef Det	DF Adj r2	Fit Std Err
0.4561028571	0.4391060714	0.4238666631

Source	Sum of Squares	DF	Mean Square	F Statistic	P>F
Regr	19.435434	3	6.4784781	36.0591	0.00000
Error	23.17652	129	0.17966295		

Description: H cronosere

X Variable: Time after initiation (TAI yr)

Xmin:	0.0000000000	Xmax:	41138.000000	Xrange:	41138.000000
Xmean:	25563.609023	Xstd:	9209.0461001	Xmedian:	27522.000000
X@Ymin:	7490.0000000	X@Ymax:	29851.000000	X@Yrange:	22361.000000

Y Variable: H

Ymin:	3.4471370510	Ymax:	6.0657529350	Yrange:	2.6186158840
Ymean:	4.3901836057	Ystd:	0.5681706056	Ymedian:	4.3261805530
Y@Xmin:	3.6772303900	Y@Xmax:	4.8532866040	Y@Xrange:	1.1760562140

Date	Time	File Source
Feb 14, 2014	3:42:39 PM	CLIPBRD.PRN

Reading Figure 15B we see residual oscillations reaching maximum around 30 kyr. AI or 12 kyr. BP. This is toward the end of the warming cycle. H has maximum at 32 kyr. AI in the vicinity of the maximum temperature. The maximum is bracketed by the 1st derivative's maximum and minimum. The minimum of the 2nd derivative points down directly onto max H. The reader can get all these and much more from the graphs and the numeric table. He or she will probably conclude with me that H is sensitively responding to climate warming.

H chronosere

Rank 1 Eqn 8160 [Line Robust None, Gaussian Errors] y=a+bx

r^2=0.39182972 DF Adj r^2=0.38247325 FitStdErr=0.43510866 Fstat=84.400199

a=4.8172271

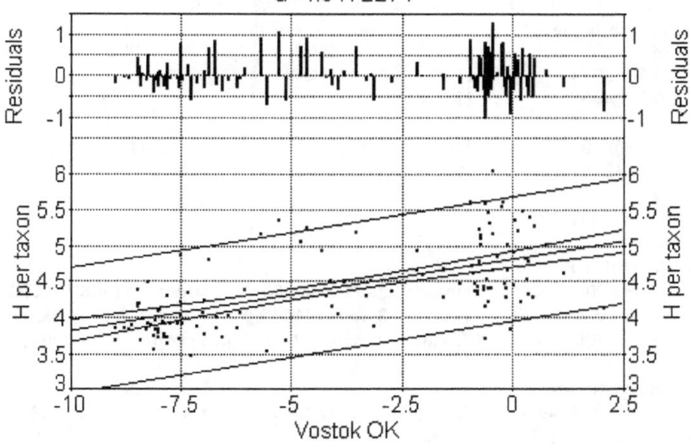

Figure 15C. H chronosere's regression on Vostok temperature differences. The regression line is within two 95% confidence belts, one for regression (inner) and the other for prediction (outer). Residuals across the top. Numeric table:

Rank 1 Eqn 8160 [Line Robust None, Gaussian Errors] y=a+bx

r^2 Coef Det	DF Adj r^2	Fit Std Err	F-value
0.3918297176	0.3824732517	0.4351086604	84.400199222

Parm	Value	Std Error	t-value	95% Confidence Limits		P>|t|
a	4.817227114	0.056661822	85.01715889	4.705136514	4.929317713	0.0000
b	0.099494119	0.010829931	9.186958105	0.078069933	0.120918305	0.0000

Area Xmin-Xmax	Area Precision		
49.538542393	1.400706e-19		
Function min	X-Value	Function max	X-Value
3.9197918834	-9.019982709	5.0221849982	2.0600000000
1st Deriv min	X-Value	1st Deriv max	X-Value
0.0994941187	-3.308675377	0.0994941187	-6.923217469
2nd Deriv min	X-Value	2nd Deriv max	X-Value
-8.83142e-10	-5.807817219	8.831422e-10	-7.297180076

Procedure	Minimization	Iterations	
LevMarqdt	Least Squares	7	
r^2 Coef Det	DF Adj r^2	Fit Std Err	r^2 Attainable
0.3918297176	0.3824732517	0.4351086604	0.9529203819

Source	Sum of Squares	DF	Mean Square	F Statistic	P>F
Regr	15.978607	1	15.978607	84.4002	0.0000
Error	24.800861	131	0.18931955		
Total	40.779468	132			
Lack Fit	22.880979	119	0.19227713	1.20181	0.3826
Pure Err	1.9198818	12	0.15999015		

Description: H chronosere

X Variable: Vostok OK

Xmin:	-9.020000000	Xmax:	2.0600000000	Xrange:	11.080000000
Xmean:	-3.903458647	Xstd:	3.4969151005	Xmedian:	-3.850000000
X@Ymin:	-7.310000000	X@Ymax:	-0.480000000	X@Yrange:	6.8300000000

Y Variable: H per taxon

Ymin:	3.4919474000	Ymax:	6.0691720600	Yrange:	2.5772246600
Ymean:	4.4288559358	Ystd:	0.5558195424	Ymedian:	4.3735789030
Y@Xmin:	3.7106399060	Y@Xmax:	4.1405788830	Y@Xrange:	0.4299389770

Date	Time	File Source
Mar 13, 2014	10:35:09 AM	CLIPBRD.PRN

Each point in Figure 15C represents one of the paleorelevés. Some interesting facts:

1. The H regression line has significant slope b indicated by parameter t for which P>|t|[31] is approaching zero. H is ascending from 4.81 nats at initiation on a 6 degree slope to 5 nats at a rate about 0.20 nats/^{O}K per taxon or 18 nats in total.

[31] Read: the probability of a t value larger than the observed t under the rule of chance is P.

2. The residuals' oscillation reaches maximum at 30 kyr. AI or 12 kyr. BP toward the end of the warming cycle. The pattern is not sharp.

Students may wish to search for other facts in Figure 15 and integrate more completely the information content of the different graphs.

H-based w_{ab} oscillations

The raw H-based instability values w_{ab} are portrayed for 133 paleorelevés in Figure 17A.

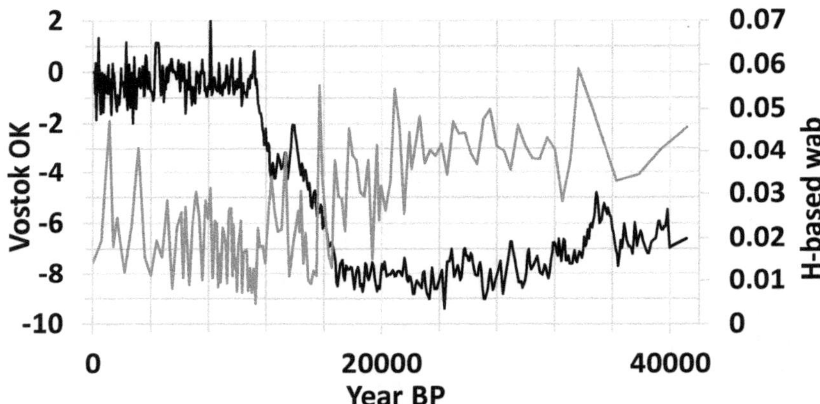

Figure 17A. H-based w_{ab} oscillations. Each of the 133 point represents one of Cwynar's paleorelevés. The Vostok graph (637 points) is included for direct reference.

The load of broad random oscillations is great, yet a trend that mirrors the temperature graph is readily recognised. Obviously this is yet another case for the Cauchy-Lorentz function. Figure 17B has the regression graphs.

H based wab chronosere
Rank 157 Eqn 8004 [Lorentzian] $y=a+b/(1+((x-c)/d)^2)$
r^2=0.48135176 DF Adj r^2=0.465144 FitStdErr=0.0099749525 Fstat=39.90783
a=0.0099635232 b=0.036957234
c=10870.189 d=13533.115

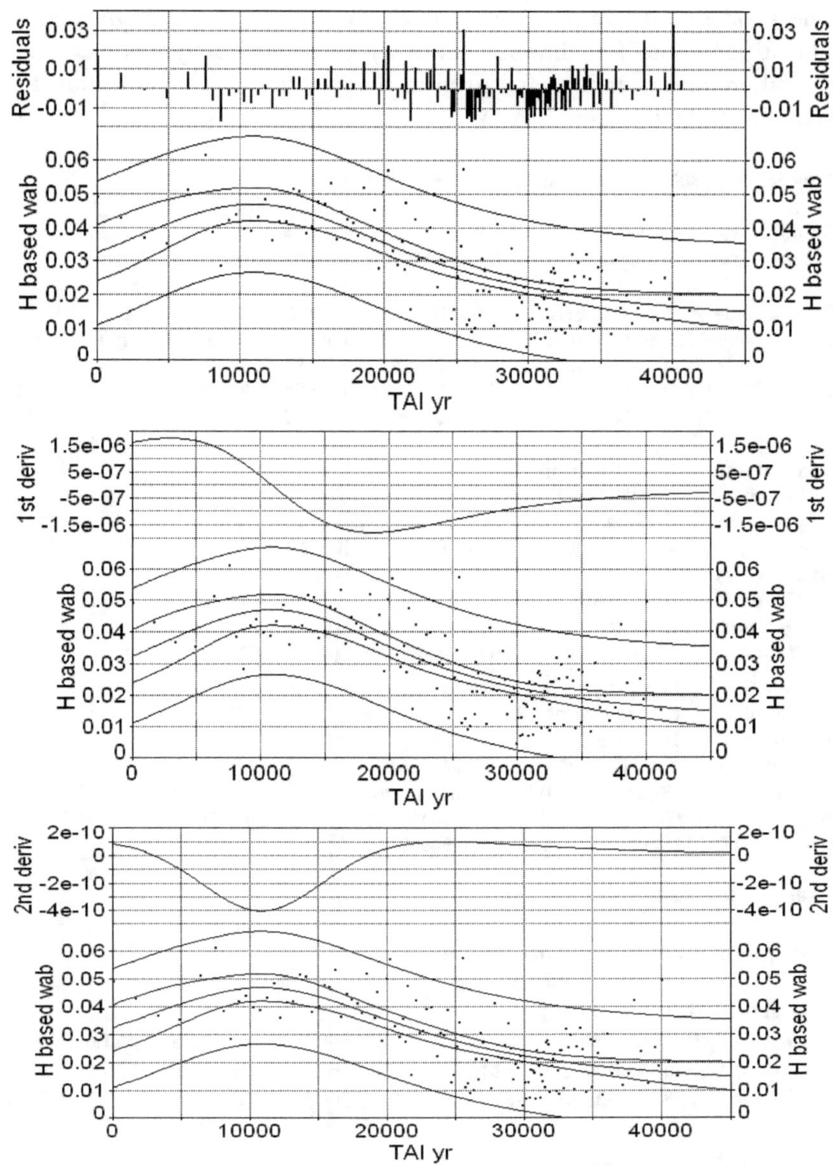

Figure 17B. Regression analysis of the instability metrics of 133 paleorelevés. *Regression line*: in centre of two 95% confidence belts, one for regression (inner) and another for prediction (outer). *Residuals, 1st and 2nd derivatives:* as identified. Note TAI time scale. The w_{ab} graph is Cauchy-Lorentz type. The differential equations are given in the chapter "Analysis of individual taxa. Numeric table:

:

Rank 157 Eqn 8004 [Lorentzian] $y=a+b/(1+((x-c)/d)^2)$

r^2 Coef Det	DF Adj r^2	Fit Std Err	F-value
0.4813517584	0.4651440008	0.0099749525	39.907829526

Parm	Value	Std Error	t-value	95% Confidence Limits		P>|t|
a	0.009963523	0.004908575	2.029819703	0.000251786	0.019675260	0.04443
b	0.036957234	0.004552345	8.118284556	0.027950308	0.045964161	0.00000
c	10870.18882	1337.126203	8.129515969	8224.651878	13515.72576	0.00000
d	13533.11475	3436.058271	3.938557988	6734.789306	20331.44020	0.00013

Area Xmin-Xmax	Area Precision		
1323.6735605	6.698883e-10		
Function min	X-Value	Function max	X-Value
0.0161207402	41138.000000	0.0469207577	10870.188763
1st Deriv min	X-Value	1st Deriv max	X-Value
-1.77375e-06	18683.540326	1.773755e-06	3056.8423291
2nd Deriv min	X-Value	2nd Deriv max	X-Value
-4.03584e-10	10870.190187	1.00896e-10	24403.300679

Procedure	Minimization	Iterations
LevMarqdt	Least Squares	29
r^2 Coef Det	DF Adj r^2	Fit Std Err
0.4813517584	0.4651440008	0.0099749525

Source	Sum of Squares	DF	Mean Square	F Statistic	P>F
Regr	0.011912448	3	0.0039708161	39.9078	0.00000
Error	0.012835458	129	9.9499677e-05		
Total	0.024747907	132			

Description: H based wab chronosere

X Variable: TAI yr

Xmin:	0.0000000000	Xmax:	41138.000000	Xrange:	41138.000000
Xmean:	25563.609023	Xstd:	9209.0461001	Xmedian:	27522.000000
X@Ymin:	29851.000000	X@Ymax:	7490.0000000	X@Yrange:	22361.000000

Y Variable: H based wab

Ymin:	0.0046312450	Ymax:	0.0616461610	Yrange:	0.0570149160
Ymean:	0.0280899833	Ystd:	0.0136924849	Ymedian:	0.0260864450
Y@Xmin:	0.0493063940	Y@Xmax:	0.0154836180	Y@Xrange:	0.0338227760

Date	**Time**	**File Source**
Feb 14, 2014	**3:13:07 PM**	**CLIPBRD.PRN**

H-based w_{ab} chronosere

Rank 1 Eqn 8160 [Line Robust None, Gaussian Errors] y=a+bx

^{2}=0.43956826 DF Adj r^2=0.43094623 FitStdErr=0.0096200315 Fstat=102.7483

a=0.017408218

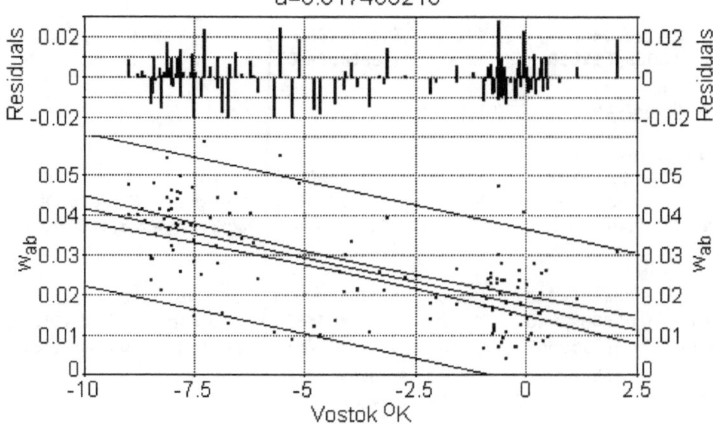

Figure 17C. Regression analysis of the H-based w_{ab} chronosere on Vostok temperature differences. The regression line is in centre within two 95% confidence belts, one for regression (inner) and the other for prediction (outer). Residuals are on top. Regression numeric table:

Rank 1 Eqn 8160 [Line Robust None, Gaussian Errors] y=a+bx

r^2 Coef Det	DF Adj r^2	Fit Std Err	F-value
0.4395682578	0.4309462310	0.0096200315	102.74835887

Parm	Value	Std Error	t-value	95% Confidence Limits		P>|t|
a	0.017408218	0.001252764	13.89584683	0.014929952	0.019886484	0.0000
b	-0.00242712	0.000239444	-10.1364865	-0.00290080	-0.00195345	0.0000

Area Xmin-Xmax	Area Precision		
0.2864690567	0.0000000000		
Function min	X-Value	Function max	X-Value
0.0124083435	2.0600000000	0.0393008305	-9.019982709
1st Deriv min	X-Value	1st Deriv max	X-Value
-0.002427124	-9.019982709	-0.002427124	-5.257344964
2nd Deriv min	X-Value	2nd Deriv max	X-Value
-3.44978e-12	-1.415257270	6.899554e-12	-6.299770446

Procedure	Minimization	Iterations	
LevMarqdt	Least Squares	7	
r^2 Coef Det	DF Adj r^2	Fit Std Err	r^2 Attainable
0.4395682578	0.4309462310	0.0096200315	0.9561798120

Source	Sum of Squares	DF	Mean Square	F Statistic	P>F
Regr	0.0095088474	1	0.0095088474	102.748	0.0000
Error	0.012123396	131	9.2545005e-05		
Total	0.021632243	132			
Lack Fit	0.011175467	119	9.3911485e-05	1.18884	0.3918
Pure Err	0.00094792896	12	7.899408e-05		

Description: H-based wab chronosere

X Variable: Vostok OK

Xmin:	-9.020000000	Xmax:	2.0600000000	Xrange:	11.080000000
Xmean:	-3.903458647	Xstd:	3.4969151005	Xmedian:	-3.850000000
X@Ymin:	-0.480000000	X@Ymax:	-7.310000000	X@Yrange:	6.8300000000

Y Variable: wab

Ymin:	0.0046154740	Ymax:	0.0590296910	Yrange:	0.0544142170
Ymean:	0.0268823945	Ystd:	0.0128015870	Ymedian:	0.0248942630
Y@Xmin:	0.0477269640	Y@Xmax:	0.0313207860	Y@Xrange:	0.0164061780

Date	Time	File Source
Mar 13, 2014	1:16:49 PM	CLIPBRD.PRN

Considering the H-based w_{ab} regression on the Vostok temperature differences the results are rather striking. Note the decline of w_{ab} on a slope of 8.4 minutes with increasing temperature. This is a rather slight decline, yet it is statistically significant. It contrasts with H which ascends on a 6 degree slope with increasing temperature. In other words stability increases (however slightly) as H increases. This contrasts with pervious results which suggest considerable destabilizing effect of climate warming.

We can take the argument further about the temperature effect on a higher level of the non-species based taxonomy, such as the functional type level of the palynomorph taxa. This topic is discussed in the next chapter.

H and w_{ab} trends within functional types

I used Raunkiaer's life form types to classify the palynomorph taxa into four broad categories:

Phanerophytes
Chamaephytes
Hemicryptophytes
Cryptophytes

I refer to these as functional type *sensu lato* to emphasise the fact that the plant traits represented point to the specific survival characteristics of plants.

The relatively low number of palynomorph taxa (89) in the data set did not support a more refined classification without critical reduction in the value of T and especially n. The four functional types are the cases or metataxa as it were. Within each metataxon the complex is a paleorelevé and the resonators are palynomorph taxa.

First the H and w_{ab} graphs are presented (Figure 18), followed by the regression analysis.

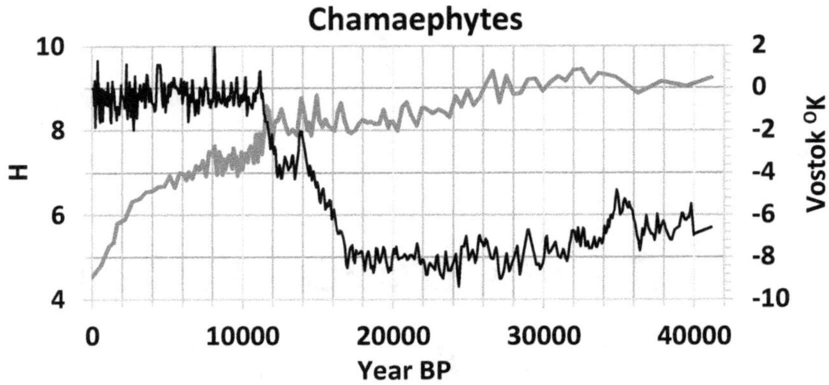
Chamaephytes
10
8
6
4
H
2
0
-2
-4
-6
-8
-10
Vostok °K
0
10000
20000
30000
40000
Year BP

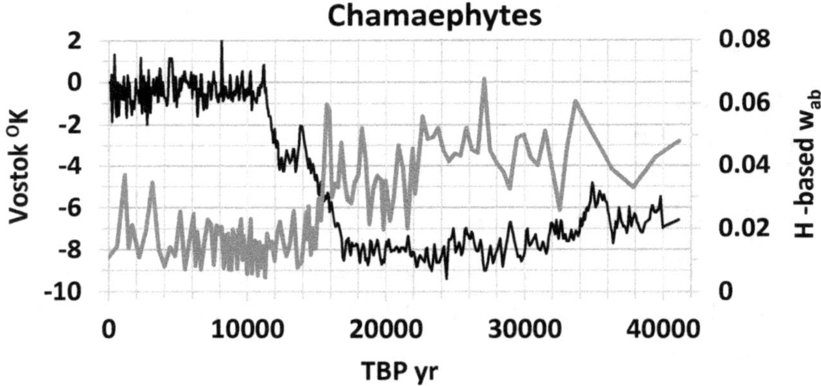
Chamaephytes
2
0
-2
-4
-6
-8
-10
Vostok °K
0.08
0.06
0.04
0.02
0
H -based w_{ab}
0
10000
20000
30000
40000
TBP yr

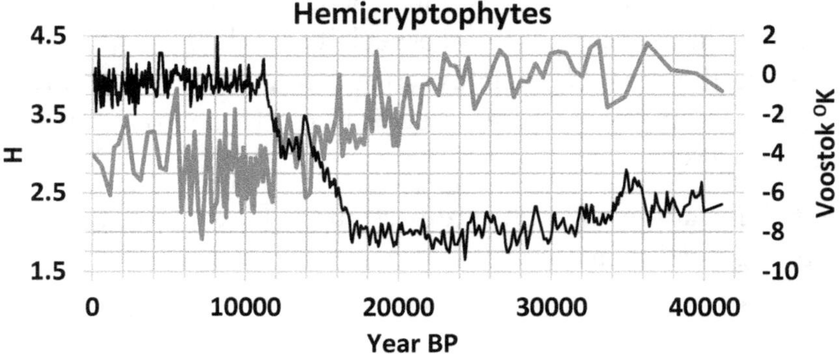
Hemicryptophytes
4.5
3.5
2.5
1.5
H
2
0
-2
-4
-6
-8
-10
Voostok °K
0
10000
20000
30000
40000
Year BP

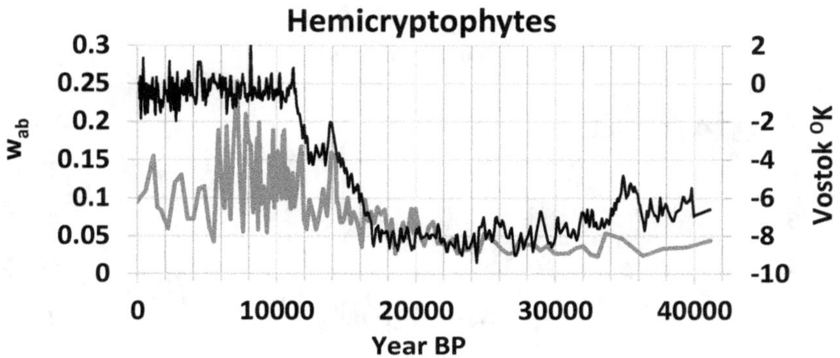
Hemicryptophytes
0.3
0.25
0.2
0.15
0.1
0.05
0
w_{ab}
2
0
-2
-4
-6
-8
-10
Vostok °K
0
10000
20000
30000
40000
Year BP

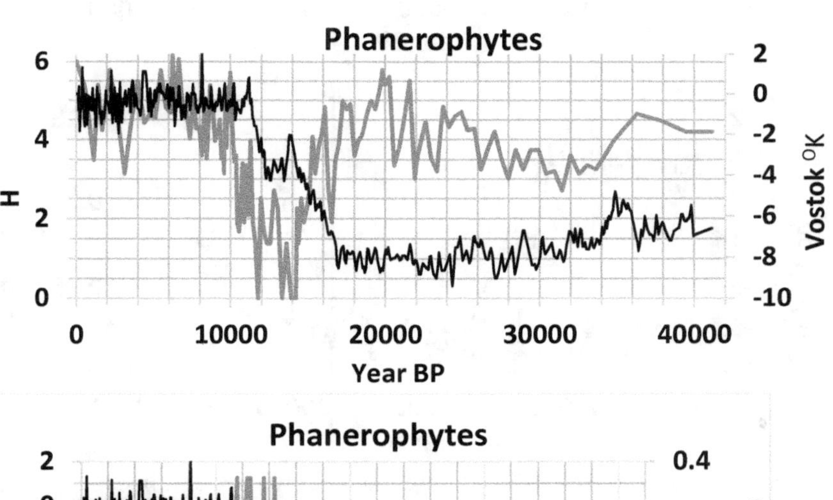
Phanerophytes
6
4
2
0
H
2
0
-2
-4
-6
-8
-10
Vostok °K
0
10000
20000
30000
40000
Year BP

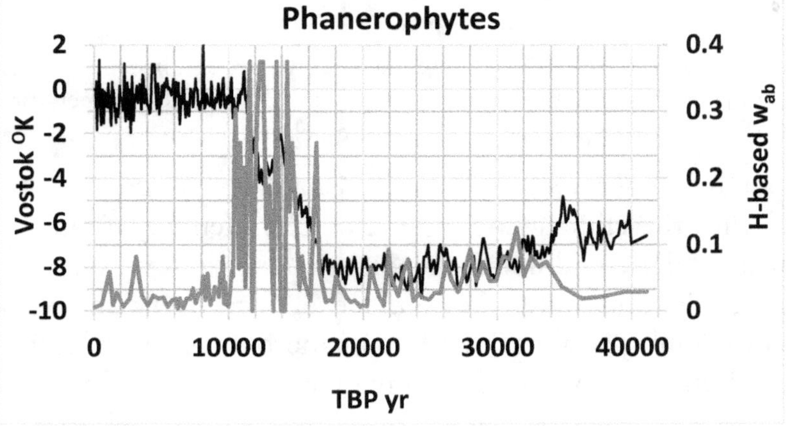
Phanerophytes
2
0
-2
-4
-6
-8
-10
Vostok °K
0.4
0.3
0.2
0.1
0
H-based w_{ab}
0
10000
20000
30000
40000
TBP yr

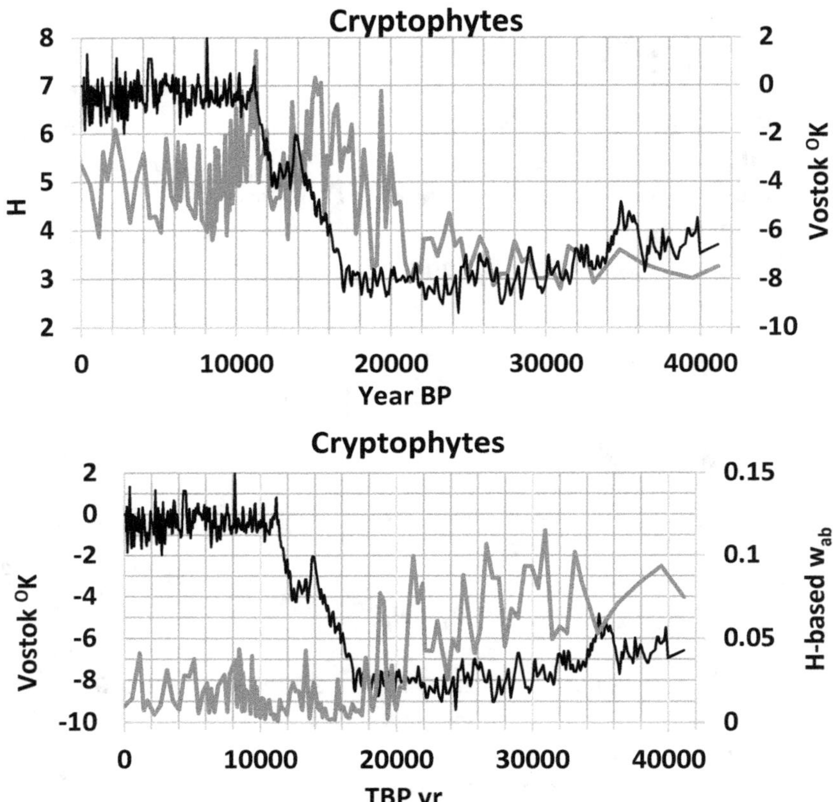

Figure 18. Graphs of the H-based potential energy state and instability level w_{ab} of 133 paleorelevés in the four life form groups. The Vostok temperature differences graph (638 point) is included for quick reference.

The potential energy state places the Chamaephyte groups' optimum into the climate cooling cycle. The Hemicryptophytes follow a similar trend but their potential energy state starts to decline earlier and undertakes some level of recovery only well within the interglacial period. The Phanerophytes have high H level on both sides of the rapid climate warming period and interestingly but understandably they have minimum in the muddle through the rapid warming process where the temperature drops back steeply by about 2 ^{O}K. The H level in the Cryptophytes is supressed during the climate cooling cycle. It reaches maximum around year 11000 BP in climate warming cycle. The instability parameter w_{ab} follows a similar pattern as H in the Chamaephytes, reverses pattern in the Hemicryptophytes, Phanerophytes and Cryptophytes. The up and down oscillations of H and w^{ab} are indicated by arrows under climate cooling and climate warming in the following table:

Functional type	H		w_{ab}	
	Cool	Warm	Cool	Warm
Chamaephytes	↑	↓	↑	↓
Hemicryptophytes	↑	↓	↓	↑
Phanerophytes	↑↓	↓↑	↓↑	↑↓
Cryptophytes	↓	↑	↑	↓

I repeat, in case the reader missed these points: The raw graphs are not convenient for input in modelling. Formal parameterised functions are needed including differential equations. We can get these in two steps.

1. Select a curve shape for which there is a function. Fit the function to the data in a high-level regression analysis.

2. Find the appropriate differential equations. For this you can use software such as David Scherfgen's Derivative Calculator which I already referenced in Footnote. Select the first and second derivative options.

To keep the coherence of the information intact we discuss all results for the Chamaephyte group in Figure 18 and then continue the presentation of other groups in a less detailed manner leaving some of the interpretations up to interested readers.

Chamaephytes

Analysis of the H chronosere

Figure 18 has the Chamaephytes' H graph. I choose the Cauchy-Lorentz equation (see caption, Figure 18) for the shape function which is already familiar to readers. So are too the differential equations. The numerical values of the equation parameters are found with the complete set of regression results in the numeric tables which follows the figures. Note the time scale: TAI = 41138 - TBP.

Chamaephytes chronosere
Rank 224 Eqn 8004 [Lorentzian] $y=a+b/(1+((x-c)/d)^2)$
r^2=0.88522709 DF Adj r^2=0.88164044 FitStdErr=0.32638 Fstat=331.65288
a=-2021.8323 b=2030.9277
c=8269.0224 d=795007.54

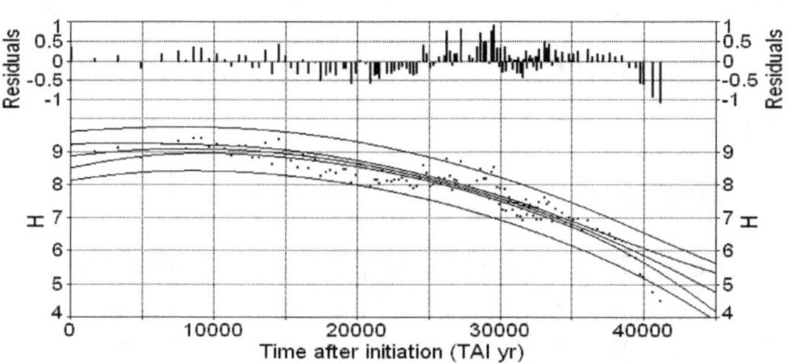

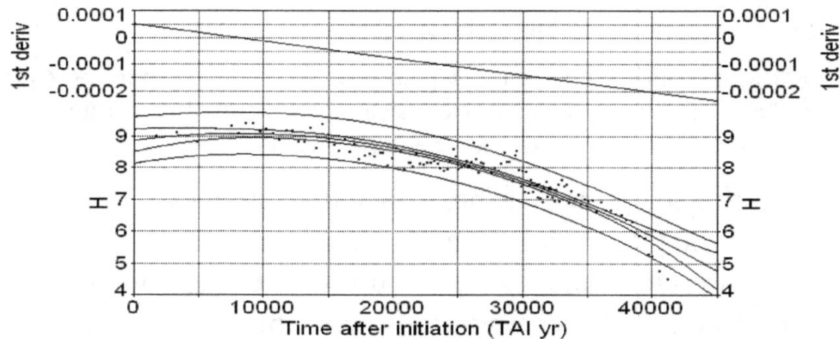

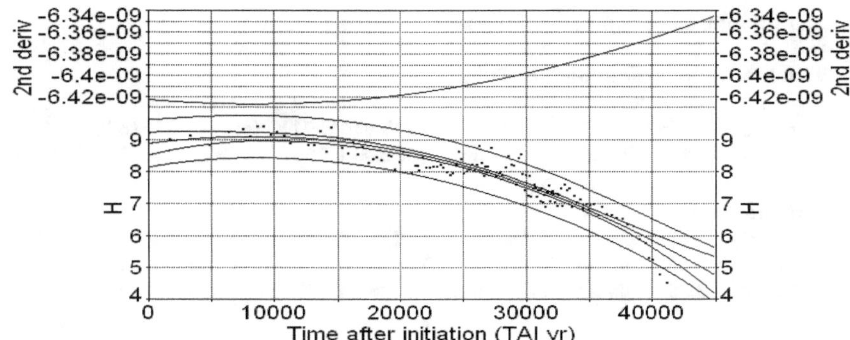

Figure 19A. Regression of H on time in the Chamaephytes' group. H graph 133 points. Each point represents a paleorelevé. *Regression line*: centre line within two 95% confidence belts, inner for regression, outer for prediction. *Residuals and derivatives:* as identified. Numeric table:

Rank 224 Eqn 8004 [Lorentzian] y=a+b/(1+((x-c)/d)²)

r^2 Coef Det	DF Adj r^2	Fit Std Err	F-value
0.8852270942	0.8816404409	0.3263800028	331.65288256

Parm	Value	Std Error	t-value	95% Confidence Limits		P>|t|
a	-2021.83226	187252.2542	-0.01079737	-372505.003	368461.3388	0.99140
b	2030.927655	187252.2487	0.010845945	-368452.233	372514.0879	0.99136
c	8269.022402	1915.542962	4.316803416	4479.073736	12058.97107	0.00003
d	795007.5425	3.67338e+07	0.021642402	-7.1884e+07	7.34737e+07	0.98277

Area Xmin-Xmax	Area Precision		
335564.40904	2.866649e-13		
Function min	X-Value	Function max	X-Value
5.6297657874	41138.000000	9.0953991878	8269.0223515
1st Deriv min	X-Value	1st Deriv max	X-Value
-0.000210516	41138.000000	5.313029e-05	1.012919e-06
2nd Deriv min	X-Value	2nd Deriv max	X-Value
-6.42665e-09	8194.1522655	-6.36094e-09	41137.288024

Procedure	Minimization	Iterations
LevMarqdt	Least Squares	200
r^2 Coef Det	DF Adj r^2	Fit Std Err
0.8852270942	0.8816404409	0.3263800028

Source	Sum of Squares	DF	Mean Square	F Statistic	P>F
Regr	105.98688	3	35.328961	331.653	0.00000
Error	13.741584	129	0.10652391		
Total	119.72847	132			

Description: Chamaephytes chronosere

X Variable: Time after initiation (TAI yr)

Xmin:	0.0000000000	Xmax:	41138.000000	Xrange:	41138.000000
Xmean:	25563.609023	Xstd:	9209.0461001	Xmedian:	27522.000000
X@Ymin:	41138.000000	X@Ymax:	8602.0000000	X@Yrange:	32536.000000

Y Variable: H

Ymin:	4.5326076720	Ymax:	9.4446992990	Yrange:	4.9120916270
Ymean:	7.8634253304	Ystd:	0.9523832371	Ymedian:	8.0047772450
Y@Xmin:	9.2279705360	Y@Xmax:	4.5326076720	Y@Xrange:	4.6953628640

Date	Time	File Source
Feb 14, 2014	11:33:04 PM	CLIPBRD.PRN

The extreme points are summarised:

Extreme points of $f(x)=\frac{b}{(x-c)^2/d^2+1}+a$	Vostok kyr. TAI or nat	Vostok kyr. TAI or ^{O}K
x of f(x) min yr.	41.1	19.5
minimum	0	-8.446
x of f(x) max yr.	8.3	34.6
maximum	9.01	-0.068
x of 1st deriv min yr.	41.1	14.1 or 37.0
x of 1st deriv max yr.	0	26.7
x of 2nd deriv min yr.	8.2	29.8
x of 2nd deriv max yr.	41.1	23.6

What can we read from the above?

1. The Cauchy-Lorentz function captures the trend in a closely perfect manner (R^2=0.88).

2. Break points of the residuals' pattern transitions:

TBP kyr.	3	8	11	16	27
TAI kyr.	38	33	30	15	14
Vostok f(x) extr point kyr.	37	35	30	14	14
	1st max	f(x) max	2nd min	1st min	1st min

The similarity of break points of H with break points of the Vostok temperature chronosere are unmistakable.

The next step is regression analysis of the 133 valued H chronosere of the Chamaephyte group and the Vostok temperature differences chronosere of nominally matching points. The graphs are in Figure 19B.

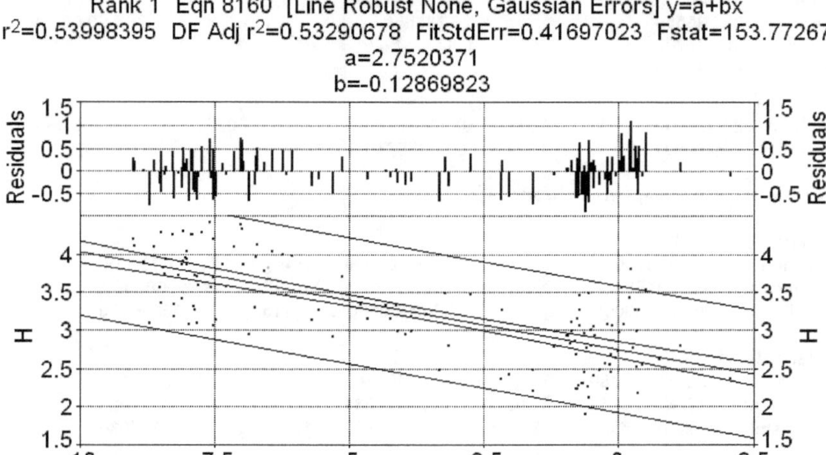

Figure 19B. Regression of H on Vostok temperature differences in the Chamaephytes' group. Number of points 133 points. Each point represents a paleorelevé. *Regression line*: centre line within two 95% confidence belts, inner for regression, outer for prediction. *Residuals:* as identified. Numeric table:

Rank 1 Eqn 8160 [Line Robust None, Gaussian Errors] y=a+bx

r^2 Coef Det	DF Adj r^2	Fit Std Err	F-value
0.5399839486	0.5329067786	0.4169702289	153.77267175

Parm	Value	Std Error	t-value	95% Confidence Limits		P>|t|
a	2.752037147	0.054299753	50.68231418	2.644619283	2.859455011	0.00000
b	-0.12869823	0.010378462	-12.4005109	-0.14922930	-0.10816716	0.00000

Area Xmin-Xmax	Area Precision		
35.454969422	0.0000000000		
Function min	X-Value	Function max	X-Value
2.4869187932	2.0600000000	3.9128929565	-9.019982709
1st Deriv min	X-Value	1st Deriv max	X-Value
-0.128698230	-7.813073963	-0.128698230	-8.514589038
2nd Deriv min	X-Value	2nd Deriv max	X-Value
-8.83143e-10	-8.592277282	4.415714e-10	-7.413485208

Procedure	Minimization	Iterations	
LevMarqdt	Least Squares	6	
r^2 Coef Det	DF Adj r^2	Fit Std Err	r^2 Attainable
0.5399839486	0.5329067786	0.4169702289	0.9727682295

Source	Sum of Squares	DF	Mean Square	F Statistic	P>F
Regr	26.735558	1	26.735558	153.773	0.00000
Error	22.776207	131	0.17386417		
Total	49.511765	132			
Lack Fit	21.427913	119	0.1800665	1.60262	0.18167
Pure Err	1.348293	12	0.11235775		

Description: Chamaephytes chronosere

X Variable: Vostok oK

Xmin:	-9.020000000	Xmax:	2.0600000000	Xrange:	11.080000000
Xmean:	-3.903458647	Xstd:	3.4969151005	Xmedian:	-3.850000000
X@Ymin:	-0.630000000	X@Ymax:	-7.600000000	X@Yrange:	6.9700000000

Y Variable: H

Ymin:	1.9095425050	Ymax:	4.4362455400	Yrange:	2.5267030350
Ymean:	3.2544053658	Ystd:	0.6124452031	Ymedian:	3.1924226020
Y@Xmin:	4.2216625990	Y@Xmax:	2.3836778960	Y@Xrange:	1.8379847030

Date	Time	File Source
Feb 14, 2014	11:39:22 PM	CLIPBRD.PRN

The linear regression line in Figure 19B traces the line of maximum separation of two groups of paleorelevés. The break points are clearly identified by the spatial pattern of the residuals:

Group 1: -8 ^{O}K, 3.8 nats
Group 2: -0.4 ^{O}K, 2.7 nats

We can do a quick test on the two groups' separation. For the temperature differences we have the standard normal variate 7.6/3.5=2.17. This indicates a low probability separation under the rule of chance. For the potential energy difference H= 1.1 nats we have $P=e^{-1.1}=0.332$. This is far too large to be considered significant.

Analysis of the Chamaephytes' w_{ab} *chronosere*

I included the w_{ab} chronosere's oscillogram and Vostok temperature graph in Figure 18. The next sets of graphs portray results from regression analysis for w_{ab} on time. The Cauchy-Lorentz function is applied. Note the time scale TAI=abs(TBP-41138).

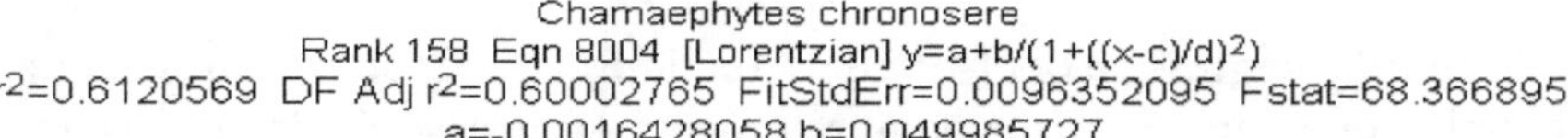

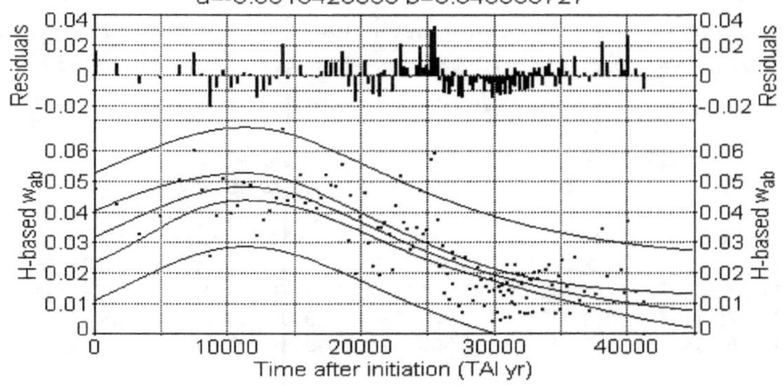

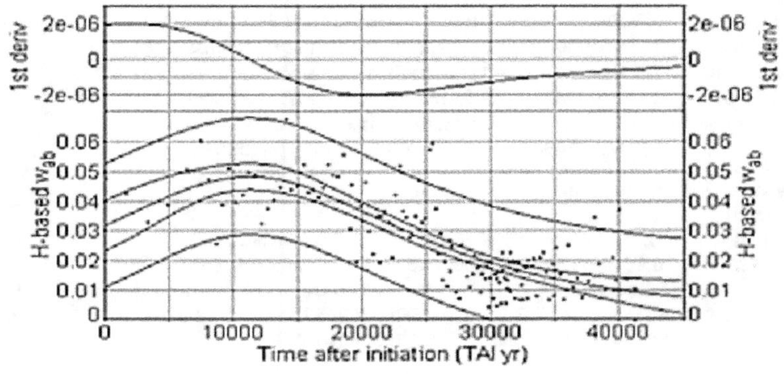

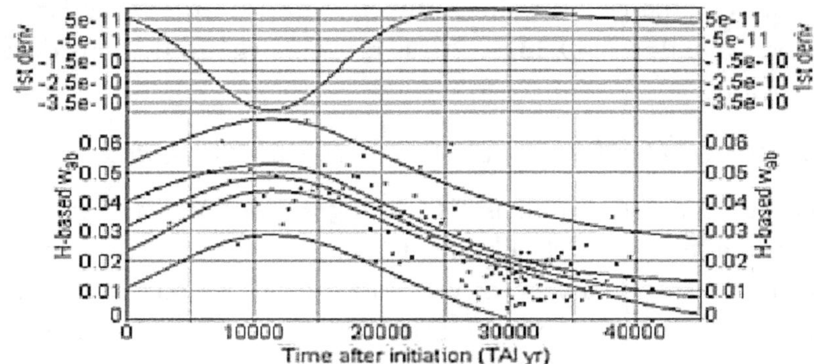

Figure 19C. Regression of H-based w_{ab} on time in the Chamaephytes' group. Number of points 133. Each point represents a paleorelevé. *Regression line:* imbedded within two 95% confidence belts, inner for regression, outer for prediction. *Residuals and derivatives:* as identified. Numeric table:

Rank 158 Eqn 8004 [Lorentzian] $y=a+b/(1+((x-c)/d)^2)$

r^2 Coef Det	DF Adj r^2	Fit Std Err	F-value
0.6120568962	0.6000276527	0.0096352095	68.366895154

Parm	Value	Std Error	t-value	95% Confidence Limits		P>|t|
a	-0.00164281	0.006495768	-0.25290401	-0.01449391	0.011208294	0.8007
b	0.049985727	0.005942733	8.411235404	0.038228740	0.061742714	0.0000
c	11247.18454	1131.203362	9.942672479	9009.233973	13485.13511	0.0000
d	16056.76534	3340.443303	4.806776792	9448.097764	22665.43292	0.0000

Area Xmin-Xmax	Area Precision		
1287.9333248	3.59597e-11		
Function min	X-Value	Function max	X-Value
0.0095510823	41138.000000	0.0483429211	11247.184623
1st Deriv min	X-Value	1st Deriv max	X-Value
-2.02195e-06	20569.039677	2.021994e-06	1976.8057899
2nd Deriv min	X-Value	2nd Deriv max	X-Value
-3.87757e-10	11247.191498	9.69393e-11	27304.292782

Procedure	Minimization	Iterations	
LevMarqdt	Least Squares	11	
r^2 Coef Det	DF Adj r^2	Fit Std Err	r^2 Attainable
0.6120568962	0.6000276527	0.0096352095	0.9982293553

Source	Sum of Squares	DF	Mean Square	F Statistic	P>F
Regr	0.019040986	3	0.0063469953	68.3669	0.0000
Error	0.012068844	130	9.2837261e-05		
Total	0.03110983	133			
Lack Fit	0.012013759	129	9.3129919e-05	1.69068	0.5567
Pure Err	5.5084456e-05	1	5.5084456e-05		

Description: Chamaephytes chronosere

X Variable: Time after initiation (TAI yr)

Xmin:	0.0000000000	Xmax:	41138.000000	Xrange:	41138.000000
Xmean:	25679.835821	Xstd:	9272.4888023	Xmedian:	27663.000000
X@Ymin:	41138.000000	X@Ymax:	14063.000000	X@Yrange:	27075.000000

Y Variable: H-based wab

Ymin:	0.0003143470	Ymax:	0.0676418600	Yrange:	0.0673275130
Ymean:	0.0263701976	Ystd:	0.0152940673	Ymedian:	0.0217930160
Y@Xmin:	0.0479980970	Y@Xmax:	0.0003143470	Y@Xrange:	0.0476837500

Date	Time	File Source
Mar 14, 2014	3:44:42 PM	CLIPBRD.PRN

The extreme points are summarised below:

Extreme points of $f(x)=\frac{b}{(x-c)^2/d^2+1}+a$	Vostok kyr. TAI or w_{ab}	Vostok kyr. TAI or ^{O}K
x of f(x) min yr.	41.1	19.5
minimum	0.0096	-8.446
x of f(x) max yr.	11.2	34.6
maximum w_{ab}	0.0483	-0.068
x of 1st deriv min yr.	20.6	14.1 or 37.0
x of 1st deriv max yr.	2.0	26.7
x of 2nd deriv min yr.	11.2	29.8
x of 2nd deriv max yr.	27.3	23.6

What is it we can read from the above:

1. The Cauchy-Lorentz function captures the trend effectively (R^2=0.61).

2. Break points of the residuals' pattern transitions:

H-based w_{ab} residuals break points					
	kyr. BP	34	18	15	4
	kyr. AI	7	**23**	**26**	**37**
Vostok f(x)		1st deriv min	2nd deriv max	1st derive max	f(x) max
Extreme Points kyr. AI		14	**24**	**27**	**35**

The break points are in three cases common time points with the Vostok extreme points. This suggests that w_{ab} is synchronous with climate change. The synchrony is such that instability is higher in the climate cooling cycle and stability increases with climate warming. The difference is minor considering that w_{ab} can have values anywhere between 0 and 0.5.

The next step is another regression analysis of the 133 valued chronosere of w_{ab} on the Vostok temperature differences. The graphs are in Figure 19D.

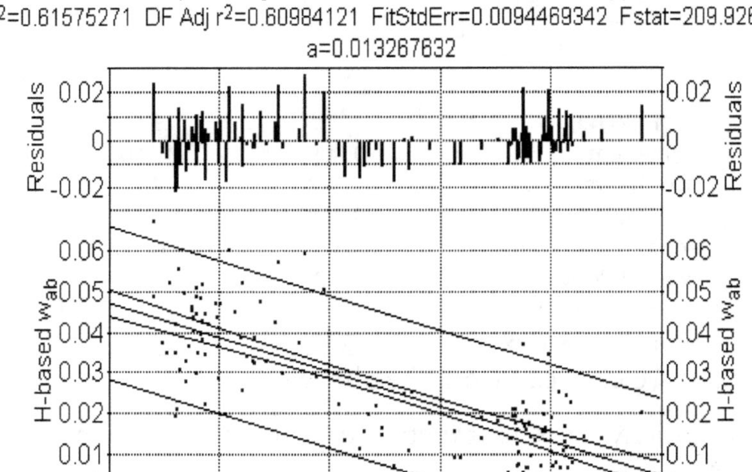

Figure 19D. Regression of H-based w_{ab} on Vostok temperature differences in the Chamaephytes' group. Number of points 133 points. Each point represents a paleorelevé. *Regression line*: centre line within two 95% confidence belts, inner for regression, outer for prediction. *Residuals:* as identified. Numeric table:

Rank 1 Eqn 8160 [Line Robust None, Gaussian Errors] y=a+bx

r^2 Coef Det	DF Adj r^2	Fit Std Err	F-value
0.6157527110	0.6098412143	0.0094469342	209.92628303

Parm	Value	Std Error	t-value	95% Confidence Limits		P>|t|
a	0.013267632	0.001230223	10.78474080	0.010833958	0.015701305	0.0000
b	-0.00340684	0.000235136	-14.4888330	-0.00387200	-0.00294169	0.0000

Area Xmin-Xmax	Area Precision		
0.2783678118	0.0000000000		
Function min	X-Value	Function max	X-Value
0.0062495331	2.0600000000	0.0439973054	-9.019982709
1st Deriv min	X-Value	1st Deriv max	X-Value
-0.003406844	-7.424060532	-0.003406844	-8.686448756
2nd Deriv min	X-Value	2nd Deriv max	X-Value
-6.89955e-12	-5.802452088	6.899554e-12	-6.804000000

Procedure	Minimization	Iterations	
LevMarqdt	Least Squares	7	
r^2 Coef Det	DF Adj r^2	Fit Std Err	r^2 Attainable
0.6157527110	0.6098412143	0.0094469342	0.9765750375

Source	Sum of Squares	DF	Mean Square	F Statistic	P>F
Regr	0.01873478	1	0.01873478	209.926	0.000C
Error	0.011691038	131	8.9244565e-05		
Total	0.030425818	132			
Lack Fit	0.010978314	119	9.2254743e-05	1.55328	0.199C
Pure Err	0.00071272365	12	5.9393637e-05		

Description: Chamaephytes chronosere

X Variable: Vostok OK

Xmin:	-9.020000000	Xmax:	2.0600000000	Xrange:	11.080000000
Xmean:	-3.903458647	Xstd:	3.4969151005	Xmedian:	-3.850000000
X@Ymin:	-0.480000000	X@Ymax:	-9.020000000	X@Yrange:	8.5400000000

Y Variable: H-based wab

Ymin:	0.0044808210	Ymax:	0.0676418600	Yrange:	0.0631610390
Ymean:	0.0265661063	Ystd:	0.0151821810	Ymedian:	0.0221586670
Y@Xmin:	0.0676418600	Y@Xmax:	0.0205636100	Y@Xrange:	0.0470782500

Date	Time	File Source
Mar 16, 2014	4:04:01 AM	CLIPBRD.PRN

The linear regression of w_{ab} in Figure 19D is similar to H in Figure 19B in that it traces the line of maximum separation of two paleorelevé groups. The group break points are clearly identified by the spatial pattern of the residuals. The group centroids have co-ordinates -8 ^{O}K, 0.04 (unitless and -0.6 ^{O}K, 0.015). We can do a quick test on the groups' separation. For the temperature differences we have the standard normal variate 7.4/3.5 =2.11. This indicates a low probability (highly significant) separation under the rue of chance. For the instability coefficient 0.025/0.015=1.67. This is on the limit of being judged significant.

From here on, with the pattern of interpretations already established, the analyses of what the graphs represent is largely the interested reader's responsibility.

Hemicryptophytes

Analysis of the H *chronosere*

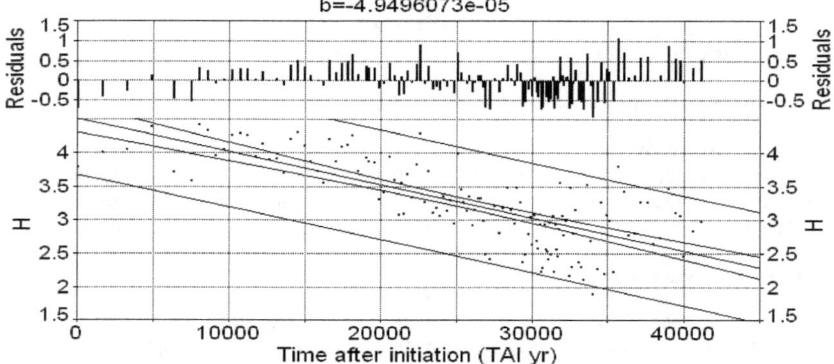

Figure 20A. Regression of H on time in the Hemicryptophytes' group. H graph 133 points. Each point represents a paleorelevé. *Regression line*: centre line within two 95% confidence belts, inner for regression, outer for prediction. *Residuals:* as identified. Numeric table:

Rank 1 Eqn 8160 [Line Robust None, Gaussian Errors] y=a+bx

r^2 Coef Det	DF Adj r^2	Fit Std Err	F-value
0.5539062951	0.5470433150	0.4106119671	162.66027489

Parm	Value	Std Error	t-value	95% Confidence Limits		P>|t|
a	4.519703616	0.105404746	42.87950773	4.311187886	4.728219347	0.00000
b	-4.9496e-05	3.88088e-06	-12.7538337	-5.7173e-05	-4.1819e-05	0.00000

Area Xmin-Xmax	Area Precision		
144049.59819	1.97305e-19		
Function min	X-Value	Function max	X-Value
2.4835341779	41138.000000	4.5197036163	1.733943e-10
1st Deriv min	X-Value	1st Deriv max	X-Value
-4.94961e-05	8838.9447775	-4.94961e-05	1573.8210730
2nd Deriv min	X-Value	2nd Deriv max	X-Value
-6.40655e-17	9123.8006158	6.406546e-17	8112.9290768

Procedure	Minimization	Iterations
LevMarqdt	Least Squares	7
r^2 Coef Det	DF Adj r^2	Fit Std Err
0.5539062951	0.5470433150	0.4106119671

Source	Sum of Squares	DF	Mean Square	F Statistic	P>F
Regr	27.424878	1	27.424878	162.66	0.00000
Error	22.086887	131	0.16860219		
Total	49.511765	132			

Description: Hemicryptophytes chronosere

X Variable: Time after initiation (TAI yr)

Xmin:	0.0000000000	Xmax:	41138.000000	Xrange:	41138.000000
Xmean:	25563.609023	Xstd:	9209.0461001	Xmedian:	27522.000000
X@Ymin:	33997.000000	X@Ymax:	8051.0000000	X@Yrange:	25946.000000

Y Variable: H

Ymin:	1.9095425050	Ymax:	4.4362455400	Yrange:	2.5267030350
Ymean:	3.2544053658	Ystd:	0.6124452031	Ymedian:	3.1924226020
Y@Xmin:	3.7956003470	Y@Xmax:	2.9916542880	Y@Xrange:	0.8039460590

Date	Time	File Source
Feb 15, 2014	8:05:15 AM	CLIPBRD.PRN

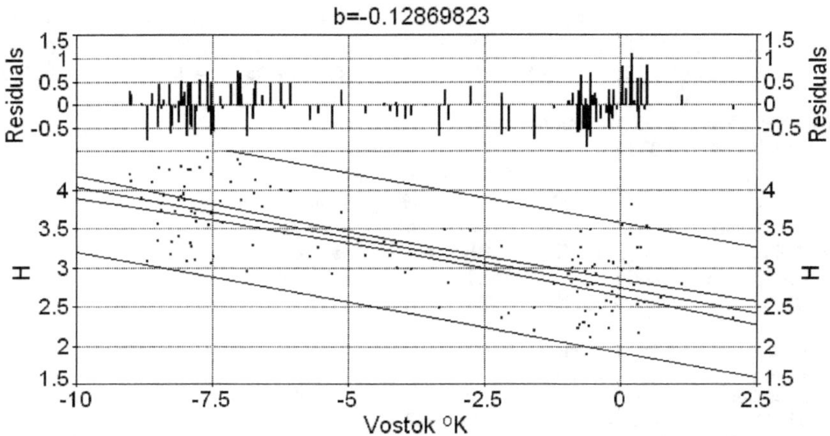

Figure 20B. Regression of H on Vostok temperature differences in the Hemicryptophytes' group. Number of points 133. Each point represents a paleo-relevé. *Regression line*: centre line within two 95% confidence belts, inner for regression, outer for prediction. *Residuals:* as identified. Numerical table:

Rank 1 Eqn 8160 [Line Robust None, Gaussian Errors] y=a+bx

r^2 Coef Det	DF Adj r^2	Fit Std Err	F-value
0.5399839486	0.5329067786	0.4169702289	153.77267175

Parm	Value	Std Error	t-value	95% Confidence Limits		P>|t|
a	2.752037147	0.054299753	50.68231418	2.644619283	2.859455011	0.00000
b	-0.12869823	0.010378462	-12.4005109	-0.14922930	-0.10816716	0.00000

Area Xmin-Xmax	Area Precision		
35.454969422	0.0000000000		
Function min	X-Value	Function max	X-Value
2.4869187932	2.0600000000	3.9128929565	-9.019982709
1st Deriv min	X-Value	1st Deriv max	X-Value
-0.128698230	-7.813073963	-0.128698230	-8.514589038
2nd Deriv min	X-Value	2nd Deriv max	X-Value
-8.83143e-10	-8.592277282	4.415714e-10	-7.413485208

Procedure	Minimization	Iterations	
LevMarqdt	Least Squares	6	
r^2 Coef Det	DF Adj r^2	Fit Std Err	r^2 Attainable
0.5399839486	0.5329067786	0.4169702289	0.9727682295

Source	Sum of Squares	DF	Mean Square	F Statistic	P>F
Regr	26.735558	1	26.735558	153.773	0.00000
Error	22.776207	131	0.17386417		
Total	49.511765	132			
Lack Fit	21.427913	119	0.1800665	1.60262	0.18167
Pure Err	1.348293	12	0.11235775		

Description: Hemicryptophyte chronosere

X Variable: Vostok oK

Xmin:	-9.020000000	Xmax:	2.0600000000	Xrange:	11.080000000
Xmean:	-3.903458647	Xstd:	3.4969151005	Xmedian:	-3.850000000
X@Ymin:	-0.630000000	X@Ymax:	-7.600000000	X@Yrange:	6.9700000000

Y Variable: H

Ymin:	1.9095425050	Ymax:	4.4362455400	Yrange:	2.5267030350
Ymean:	3.2544053658	Ystd:	0.6124452031	Ymedian:	3.1924226020
Y@Xmin:	4.2216625990	Y@Xmax:	2.3836778960	Y@Xrange:	1.8379847030

Date	Time	File Source
Feb 15, 2014	8:13:31 AM	CLIPBRD.PRN

Analysis of the Hemicryptophytes' w_{ab} *chronosere*

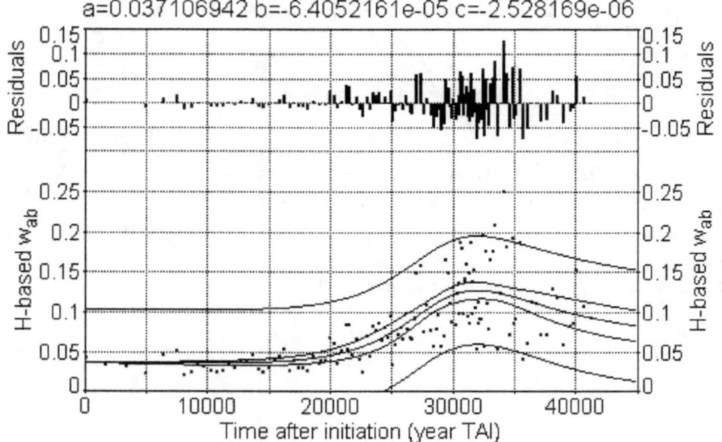

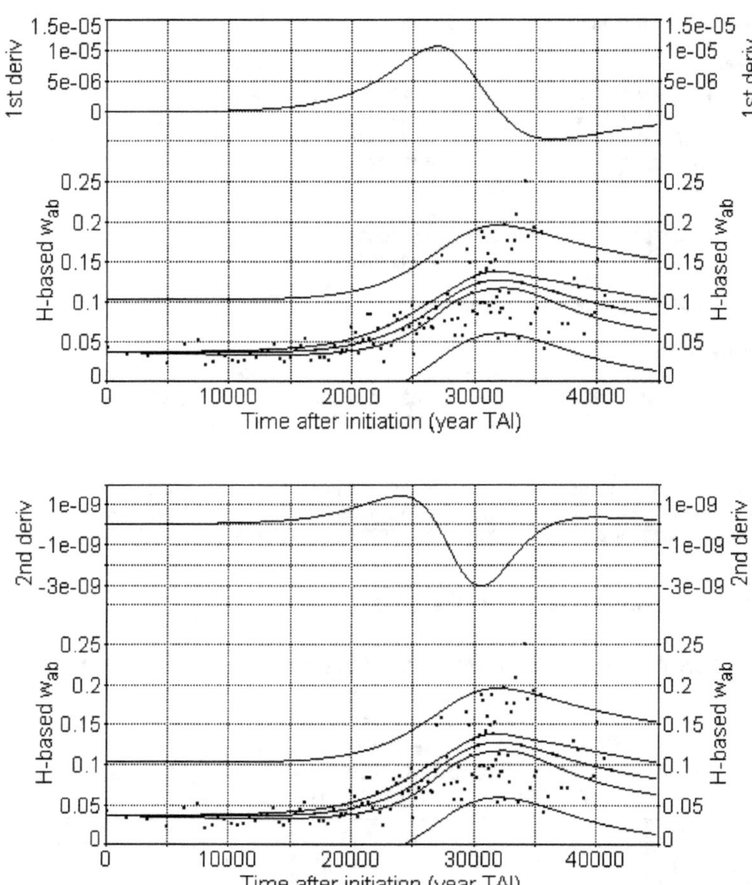

Figure 20C. Regression of H-based w_{ab} on time in the Hemicryptophytes' group. Number of points 133. Each point represents a paleorelevé. *Regression line*: centre line within two 95% confidence belts, inner for regression, outer for prediction. *Residuals and derivatives:* as identified. Numeric table:

Rank 63 Eqn 7903 $y=(a+cx+ex^2)/(1+bx+dx^2)$ [NL]

r^2 Coef Det	DF Adj r^2	Fit Std Err	F-value
0.5389805976	0.5208302274	0.0339715597	37.411395335

Parm	Value	Std Error	t-value	95% Confidence Limits		P>|t|
a	0.037106942	6.57867e-12	5.6405e+09	0.037106942	0.037106942	0.0000
b	-6.4052e-05	9.18162e-09	-6976.12800	-6.407e-05	-6.4034e-05	0.0000
c	-2.5282e-06	9.875e-08	-25.6017082	-2.7236e-06	-2.3328e-06	0.0000
d	1.09386e-09	2.32346e-11	47.07894186	1.04789e-09	1.13984e-09	0.0000
e	5.15729e-11	5.39047e-12	9.567421279	4.09069e-11	6.22388e-11	0.0000

Area Xmin-Xmax	Area Precision		
2820.4769194	4.278156e-12		
Function min	X-Value	Function max	X-Value
0.0361824542	8779.3735914	0.1279929979	32058.237315
1st Deriv min	X-Value	1st Deriv max	X-Value
-4.64665e-06	36436.580873	1.071656e-05	27038.723321
2nd Deriv min	X-Value	2nd Deriv max	X-Value
-3.02647e-09	30623.620638	1.458414e-09	24013.246919

Singularities [Data Range]

None

Singularities [All Other]

None

Soln Vector	Covar Matrix	SVD Cond
LvMrq/SVD	SVDecomp	1.029461e+21
r^2 Coef Det	DF Adj r^2	Fit Std Err
0.5389805976	0.5208302274	0.0339715597

Source	Sum of Squares	DF	Mean Square	F Statistic	P>F
Regr	0.17270101	4	0.043175252	37.4114	0.0000
Error	0.14772056	128	0.0011540669		
Total	0.32042157	132			

Description: Hemicryptophyte chronosere

X Variable: Time after initiation (year TAI)

Xmin:	0.0000000000	Xmax:	41138.000000	Xrange:	41138.000000
Xmean:	25563.609023	Xstd:	9209.0461001	Xmedian:	27522.000000
X@Ymin:	8051.0000000	X@Ymax:	33997.000000	X@Yrange:	25946.000000

Y Variable: H-based wab

Ymin:	0.0234002320	Ymax:	0.2524005490	Yrange:	0.2290003170
Ymean:	0.0865970218	Ystd:	0.0492690177	Ymedian:	0.0787706380
Y@Xmin:	0.0439290760	Y@Xmax:	0.0953676850	Y@Xrange:	0.0514386090

Date	Time	File Source
Apr 1, 2014	10:39:49 AM	CLIPBRD.PRN

Hemicryptophytes chronosere

Rank 1 Eqn 8160 [Line Robust None, Gaussian Errors] y=a+bx

r^2=0.4708301 DF Adj r^2=0.46268903 FitStdErr=0.035976821 Fstat=116.55754

a=0.12433429

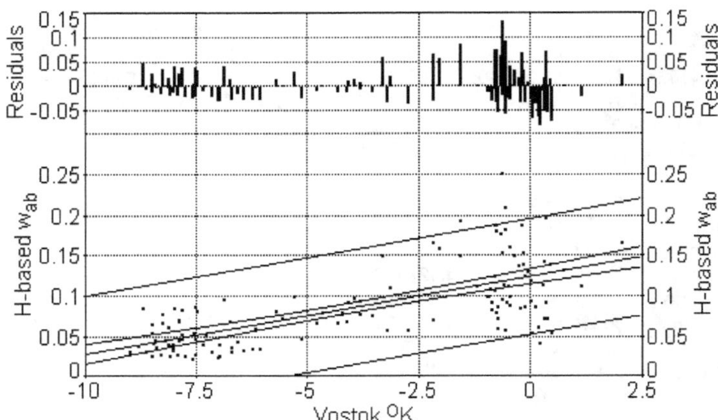

Figure 20D. Regression of H-based w_{ab} on Vostok temperature differences in the Hemicryptophytes' group. Number of points 133. Each point represents a paleorelevé. *Regression line*: centre line within two 95% confidence belts, inner for regression, outer for prediction. *Residuals:* as identified. Numeric table

Rank 1 Eqn 8160 [Line Robust None, Gaussian Errors] y=a+bx

r^2 Coef Det	DF Adj r^2	Fit Std Err	F-value
0.4708301014	0.4626890260	0.0359768212	116.55754313

Parm	Value	Std Error	t-value	95% Confidence Limits		P>|t|
a	0.124334294	0.004685065	26.53843689	0.115066119	0.133602470	0.0000
b	0.009667650	0.000895469	10.79618188	0.007896198	0.011439102	0.0000

Area Xmin-Xmax	Area Precision		
1.0048548595	0.0000000000		
Function min	X-Value	Function max	X-Value
0.0371322572	-9.019982709	0.1442496536	2.0600000000
1st Deriv min	X-Value	1st Deriv max	X-Value
0.0096676501	1.5758843978	0.0096676501	-8.333867661
2nd Deriv min	X-Value	2nd Deriv max	X-Value
-2.75982e-11	1.8122765704	2.759818e-11	0.7089166685

Procedure	Minimization	Iterations	
LevMarqdt	Least Squares	7	
r^2 Coef Det	DF Adj r^2	Fit Std Err	r^2 Attainable

0.4708301014 0.4626890260 0.0359768212 0.9813537654

Source	Sum of Squares	DF	Mean Square	F Statistic	P>F
Regr	0.15086412	1	0.15086412	116.558	0.0000
Error	0.16955745	131	0.0012943317		
Total	0.32042157	132			
Lack Fit	0.16358279	119	0.0013746453	2.76095	0.0260
Pure Err	0.0059746557	12	0.00049788798		

Description: Hemicryptophyte chronosere

X Variable: Vostok OK

Xmin:	-9.020000000	Xmax:	2.0600000000	Xrange:	11.080000000
Xmean:	-3.903458647	Xstd:	3.4969151005	Xmedian:	-3.850000000
X@Ymin:	-7.600000000	X@Ymax:	-0.630000000	X@Yrange:	6.9700000000

Y Variable: H-based wab

Ymin:	0.0234002320	Ymax:	0.2524005490	Yrange:	0.2290003170
Ymean:	0.0865970218	Ystd:	0.0492690177	Ymedian:	0.0787706380
Y@Xmin:	0.0289177880	Y@Xmax:	0.1674159550	Y@Xrange:	0.1384981670

Date	Time	File Source
Apr 1, 2014	11:22:50 AM	CLIPBRD.PRN

Phanerophytes

Analysis of the H *chronosere*

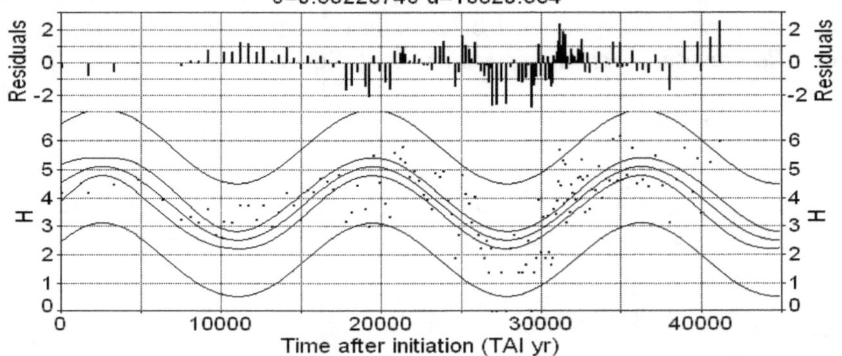

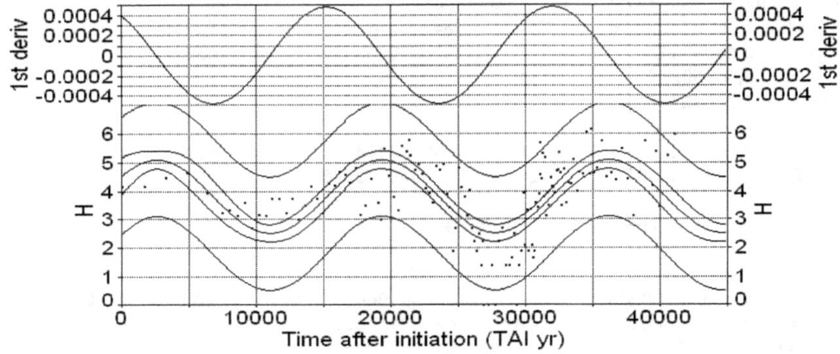

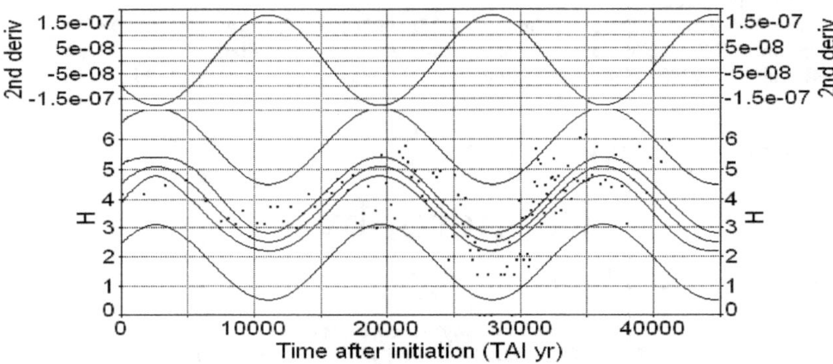

Figure 21A. Regression of H on time in the Phanerophytes' group. Number of points: 133. Each point represents a paleorelevé. *Regression line*: centre line within two 95% confidence belts, inner for regression, outer for prediction. *Residuals and derivatives:* as identified. Equations:

$$f(x) = b \cdot \sin\left(\frac{2px}{d} + c\right) + a$$

$$\frac{d}{dx} f(x) = \frac{2bp \cdot \cos\left(\frac{2px}{d} + c\right)}{d}$$

$$\frac{d^2}{d^2x} f(x) = \frac{-4bp^2 \cdot \sin\left(\frac{2px}{d} + c\right)}{d^2}$$

Numeric table:

Rank 64 Eqn 8014 [Sine] $y=a+b\sin(2\pi x/d+c)$

r^2 Coef Det	DF Adj r^2	Fit Std Err	F-value
0.4388497288	0.4213137828	0.9925798336	33.628315458

Parm	Value	Std Error	t-value	95% Confidence Limits		P>|t|
a	3.808196786	0.088020208	43.26502812	3.634046648	3.982346924	0.00000
b	1.292113315	0.128930603	10.02177361	1.037020959	1.547205670	0.00000
c	0.592267456	0.281248752	2.105849187	0.035809912	1.148725001	0.03716
d	16820.93398	456.8433768	36.81991429	15917.05817	17724.80979	0.00000

Area Xmin-Xmax	Area Precision		
162882.58291	5.23476e-18		
Function min	X-Value	Function max	X-Value
2.5160834714	11030.119296	5.1003101008	2619.6535305
1st Deriv min	X-Value	1st Deriv max	X-Value
-0.000482648	6824.8865411	0.0004826478	15235.353227
2nd Deriv min	X-Value	2nd Deriv max	X-Value
-1.80285e-07	2619.6509693	1.802852e-07	11030.108417

Procedure	Minimization	Iterations
LevMarqdt	Least Squares	13
r^2 Coef Det	DF Adj r^2	Fit Std Err
0.4388497288	0.4213137828	0.9925798336

Source	Sum of Squares	DF	Mean Square	F Statistic	P>F
Regr	99.393335	3	33.131112	33.6283	0.00000
Error	127.0927	129	0.98521473		
Total	226.48603	132			

Description: Phanerophytes chronosere

X Variable: Time after inisiciation (TAI yr)

Xmin:	0.0000000000	Xmax:	41138.000000	Xrange:	41138.000000
Xmean:	25563.609023	Xstd:	9209.0461001	Xmedian:	27522.000000
X@Ymin:	26905.000000	X@Ymax:	34897.000000	X@Yrange:	7992.0000000

Y Variable: H

Ymin:	0.0000000000	Ymax:	6.1676376900	Yrange:	6.1676376900
Ymean:	3.7778258207	Ystd:	1.3098867475	Ymedian:	3.9703048670
Y@Xmin:	4.1986036930	Y@Xmax:	6.0072945290	Y@Xrange:	1.8086908360

Date	Time	File Source
Feb 15, 2014	10:27:34 AM	CLIPBRD.PRN

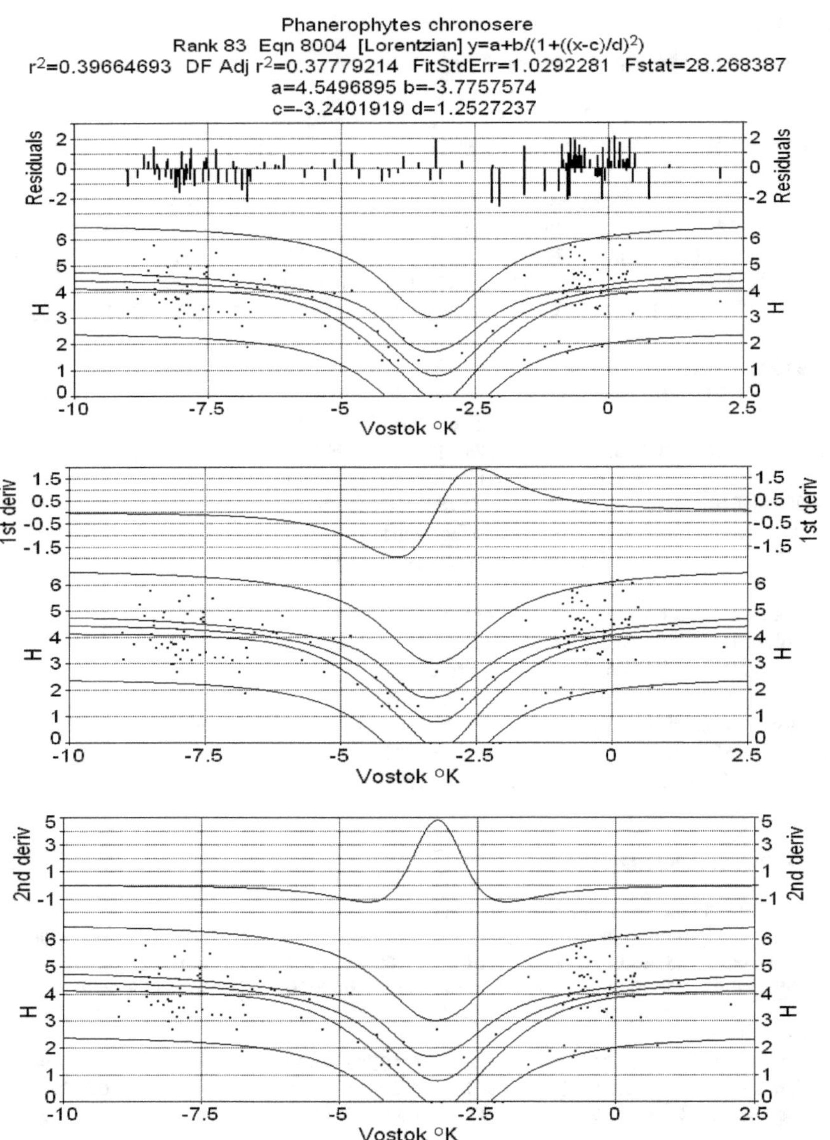

Figure 21B. Regression of H on Vostok temperature differences in the Phanerophytes' group. Number of points 133. Each point represents a paleorelevé. *Regression line*: centre line within two 95% confidence belt, inner for regression, outer for prediction. *Residuals and derivatives:* as identified. Shape and differential equations:

Rank 83 Eqn 8004 [Lorentzian] y=a+b/(1+((x-c)/d)^2)

r^2 Coef Det	DF Adj r^2	Fit Std Err	F-value
0.3966469283	0.3777921448	1.0292280715	28.268386649

Parm	Value	Std Error	t-value	95% Confidence Limits		P>|t|
a	4.549689544	0.200792911	22.65861634	4.152415851	4.946963236	0.00000
b	-3.77575739	0.456511558	-8.27089112	-4.67897669	-2.87253809	0.00000
c	-3.24019188	0.144512943	-22.4214650	-3.52611428	-2.95426948	0.00000
d	1.252723749	0.277463803	4.514908747	0.703754820	1.801692678	0.00001

Area Xmin-Xmax	Area Precision		
37.658259421	1.411713e-13		
Function min	X-Value	Function max	X-Value
0.7739321510	-3.240191691	4.3802739937	-9.019982709
1st Deriv min	X-Value	1st Deriv max	X-Value
-1.957675319	-3.963450851	1.9576753189	-2.516931501
2nd Deriv min	X-Value	2nd Deriv max	X-Value
-1.202994007	-1.987473741	4.8119760274	-3.240195377

Procedure	Minimization	Iterations	
LevMarqdt	Least Squares	18	
r^2 Coef Det	DF Adj r^2	Fit Std Err	r^2 Attainable
0.3966469283	0.3777921448	1.0292280715	0.9494809446

Source	Sum of Squares	DF	Mean Square	F Statistic	P>F
Regr	89.83499	3	29.944997	28.2684	0.00000
Error	136.65104	129	1.0593104		
Total	226.48603	132			
Lack Fit	125.20918	117	1.070164	1.12237	0.44182
Pure Err	11.441861	12	0.95348838		

Description: Phanerophytes chronosere

X Variable: Vostok oK

Xmin:	-9.020000000	Xmax:	2.0600000000	Xrange:	11.080000000
Xmean:	-3.903458647	Xstd:	3.4969151005	Xmedian:	-3.850000000
X@Ymin:	-3.340000000	X@Ymax:	0.0900000000	X@Yrange:	3.4300000000

Y Variable: H

Ymin:	0.0000000000	Ymax:	6.1676376900	Yrange:	6.1676376900
Ymean:	3.7778258207	Ystd:	1.3098867475	Ymedian:	3.9703048670
Y@Xmin:	4.1986036930	Y@Xmax:	3.6024609680	Y@Xrange:	0.5961427250

Date	Time	File Source
Feb 15, 2014	1:59:41 PM	CLIPBRD.PRN

Analysis of the Phanerophytes' w_{ab} *chronosere*

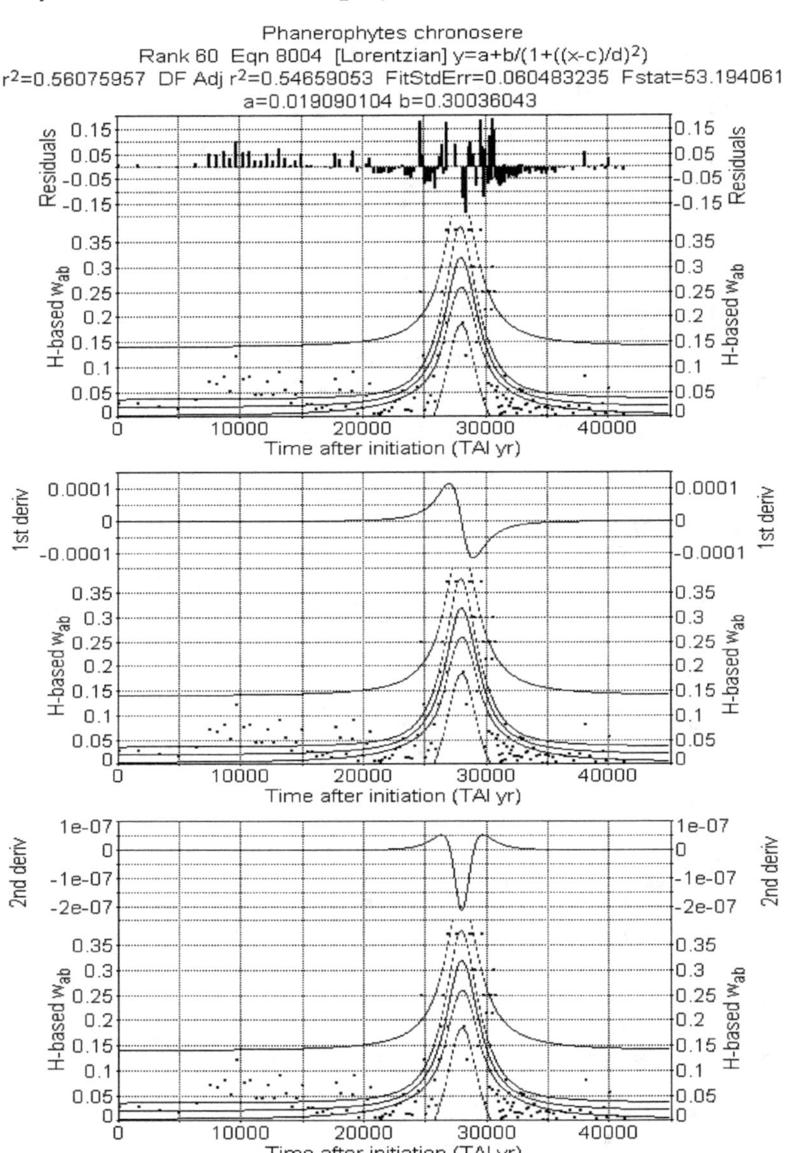

Figure 21C. Regression of H-based w_{ab} on time in the Phanerophytes' group. Number of points 133. Each point represents a paleorelevé. *Regression line*: centre line within two 95% confidence belts, inner for regression, outer for prediction. *Residuals and derivatives:* as identified. Numeric table:

Rank 60 Eqn 8004 [Lorentzian] $y=a+b/(1+((x-c)/d)^2)$

r^2 Coef Det	DF Adj r^2	Fit Std Err	F-value
0.5607595722	0.5465905261	0.0604832349	53.194061142

Parm	Value	Std Error	t-value	95% Confidence Limits		P>|t|
a	0.019090104	0.008236596	2.317717582	0.002788858	0.035391351	0.0220
b	0.300360431	0.029730224	10.10286478	0.241520628	0.359200234	0.0000
c	28100.58344	140.7731575	199.6160628	27821.97589	28379.19099	0.0000
d	1670.599482	229.5088584	7.279019615	1216.372967	2124.825998	0.0000

Area Xmin-Xmax	Area Precision		
2267.9777911	9.172357e-06		
Function min	X-Value	Function max	X-Value
0.0201479551	1.733943e-10	0.3194505353	28100.583987
1st Deriv min	X-Value	1st Deriv max	X-Value
-0.000116778	29065.108514	0.0001167783	27136.064405
2nd Deriv min	X-Value	2nd Deriv max	X-Value
7.972023e-12	0.0001429816	5.381063e-08	26429.984419
Procedure	Minimization	Iterations	
LevMarqdt	Least Squares	13	
r^2 Coef Det	DF Adj r^2	Fit Std Err	
0.5607595722	0.5465905261	0.0604832349	

Source	Sum of Squares	DF	Mean Square	F Statistic	P>F
Regr	0.58378701	3	0.19459567	53.1941	0.0000
Error	0.45727771	125	0.0036582217		
Total	1.0410647	128			

Description: Phanerophytes chronosere

X Variable: Time after initiation (TAI yr)

Xmin:	0.0000000000	Xmax:	41138.000000	Xrange:	41138.000000
Xmean:	25493.511628	Xstd:	9341.5239782	Xmedian:	27522.000000
X@Ymin:	34897.000000	X@Ymax:	26739.000000	X@Yrange:	8158.0000000

Y Variable: H-based wab

Ymin:	0.0041835760	Ymax:	0.3750000000	Yrange:	0.3708164240
Ymean:	0.0714544649	Ystd:	0.0901849107	Ymedian:	0.0370233820
Y@Xmin:	0.0295820670	Y@Xmax:	0.0049093630	Y@Xrange:	0.0246727040

Date	Time	File Source
Mar 16, 2014	3:53:12 PM	CLIPBRD.PRN

Phanerophytes chronosere

Rank 85 Eqn 8004 [Lorentzian] $y=a+b/(1+((x-c)/d)^2)$

r^2=0.44322538 DF Adj r^2=0.42526491 FitStdErr=0.068096309 Fstat=33.16912

a=0.015579103 b=0.28362402

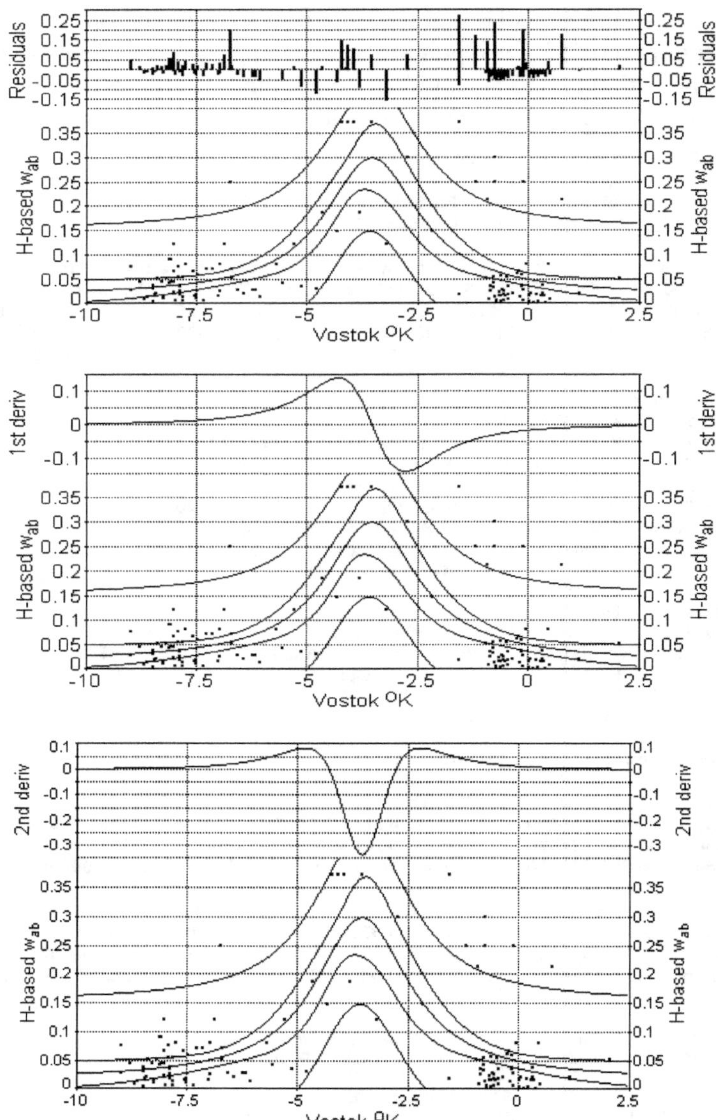

Figure 21D. Regression of H-based w_{ab} on the Vostok temperature differences in the Phanerophytes' group. Number of points 133s. Each point represents a paleorelevé. *Regression line*: centre line within two 95% confidence belts, inner for regression, outer for prediction. *Residuals and derivatives:* as identified. Numeric table:

Rank 85 Eqn 8004 [Lorentzian] $y=a+b/(1+((x-c)/d)^2)$

r^2 Coef Det	DF Adj r^2	Fit Std Err	F-value
0.4432253836	0.4252649121	0.0680963088	33.169120458

Parm	Value	Std Error	t-value	95% Confidence Limits		P>|t|
a	0.015579103	0.015466579	1.007275341	-0.01503118	0.046189382	0.3157
b	0.283624018	0.032656383	8.685102161	0.218992984	0.348255052	0.0000
c	-3.51283869	0.150728204	-23.3057822	-3.81114852	-3.21452887	0.0000
d	1.303312185	0.300633851	4.335214352	0.708320482	1.898303888	0.0000

Area Xmin-Xmax	Area Precision		
1.1630845514	1.273738e-13		
Function min	X-Value	Function max	X-Value
0.0302873286	2.0600000000	0.2992031216	-3.512837595
1st Deriv min	X-Value	1st Deriv max	X-Value
-0.141346951	-2.760370104	0.1413469511	-4.265307961
2nd Deriv min	X-Value	2nd Deriv max	X-Value
-0.333945880	-3.512839462	0.0834864701	-4.816146488

Procedure	Minimization	Iterations	
LevMarqdt	Least Squares	12	
r^2 Coef Det	DF Adj r^2	Fit Std Err	r^2 Attainable
0.4432253836	0.4252649121	0.0680963088	0.9298608295

Source	Sum of Squares	DF	Mean Square	F Statistic	P>F
Regr	0.46142631	3	0.15380877	33.1691	0.0000
Error	0.57963841	125	0.0046371073		
Total	1.0410647	128			
Lack Fit	0.50661899	113	0.0044833539	0.736794	0.8029
Pure Err	0.073019416	12	0.0060849513		

Description: Phanerophytes chronosere

X Variable: Vostok OK

Xmin:	-9.020000000	Xmax:	2.0600000000	Xrange:	11.080000000
Xmean:	-3.941007752	Xstd:	3.5430536893	Xmedian:	-3.960000000
X@Ymin:	0.0900000000	X@Ymax:	-4.110000000	X@Yrange:	4.2000000000

Y Variable: h-based wab

Ymin:	0.0041835760	Ymax:	0.3750000000	Yrange:	0.3708164240
Ymean:	0.0714544649	Ystd:	0.0901849107	Ymedian:	0.0370233820
Y@Xmin:	0.0295820670	Y@Xmax:	0.0530272840	Y@Xrange:	0.0234452170

Date	Time	File Source
Mar 16, 2014	4:29:15 PM	CLIPBRD.PRN

Cryptophytes

Analysis of the H chronosere

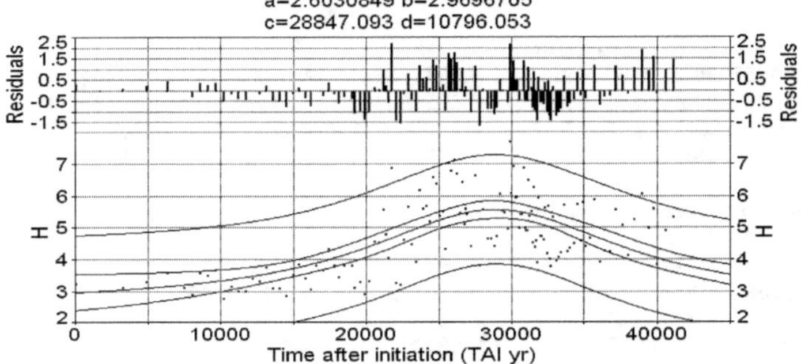

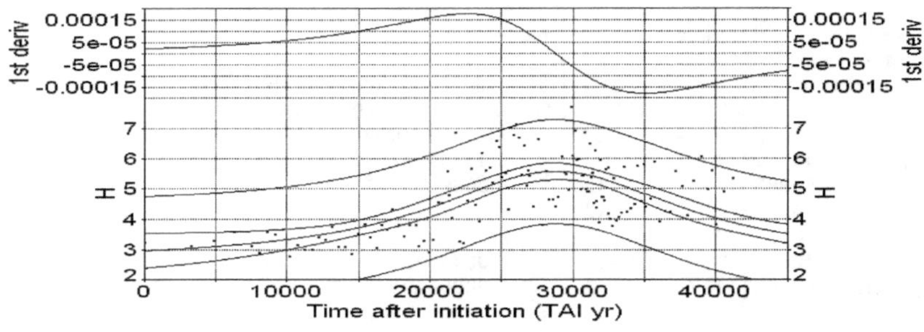

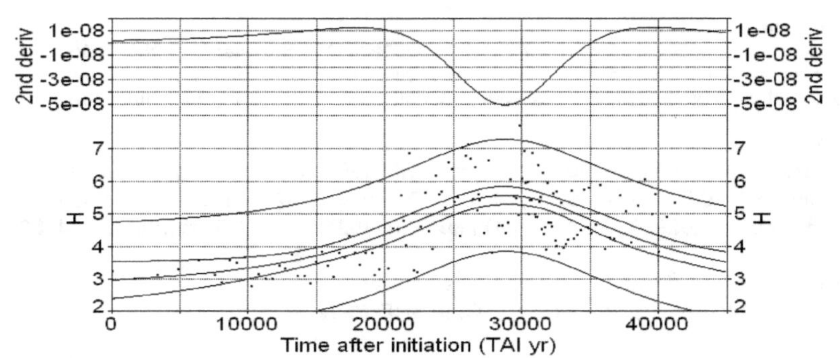

Figure 22A. Regression of H on time in the Cryptophytes' group. Number of points 133. Each point represents a paleorelevé. *Regression line*: centre line within two 95% confidence belts, inner for regression, outer for prediction. *Residuals and derivatives:* as identified. Numeric table:

Rank 115 Eqn 8004 [Lorentzian] $y=a+b/(1+((x-c)/d)^2)$

r^2 Coef Det	DF Adj r^2	Fit Std Err	F-value
0.4591213852	0.4422189285	0.8572832272	36.500277558

Parm	Value	Std Error	t-value	95% Confidence Limits		P>|t|
a	2.603084880	0.456955255	5.696585932	1.698987715	3.507182045	0.00000
b	2.969670527	0.429030213	6.921821434	2.120823742	3.818517312	0.00000
c	28847.09292	599.1307077	48.14824637	27661.69814	30032.48770	0.00000
d	10796.05301	2353.343500	4.587538117	6139.905254	15452.20076	0.00001

Area Xmin-Xmax	Area Precision		
173218.61188	2.251178e-16		
Function min	X-Value	Function max	X-Value
2.9679270427	1.733943e-10	5.5727554065	28847.093123
1st Deriv min	X-Value	1st Deriv max	X-Value
-0.000178663	35080.207900	0.0001786632	22613.993778
2nd Deriv min	X-Value	2nd Deriv max	X-Value
-5.09575e-08	28847.090040	1.273938e-08	18051.032576

Procedure	Minimization	Iterations
LevMarqdt	Least Squares	16
r^2 Coef Det	DF Adj r^2	Fit Std Err
0.4591213852	0.4422189285	0.8572832272

Source	Sum of Squares	DF	Mean Square	F Statistic	P>F
Regr	80.475943	3	26.825314	36.5003	0.00000
Error	94.806555	129	0.73493453		
Total	175.2825	132			

Description: Cryptophytes chronosere

X Variable: Time after initiation (TAI yr)

Xmin:	0.0000000000	Xmax:	41138.000000	Xrange:	41138.000000
Xmean:	25563.609023	Xstd:	9209.0461001	Xmedian:	27522.000000
X@Ymin:	10198.000000	X@Ymax:	29851.000000	X@Yrange:	19653.000000

Y Variable: H

Ymin:	2.7905996430	Ymax:	7.7236320400	Yrange:	4.9330323970
Ymean:	4.7257573906	Ystd:	1.1523444408	Ymedian:	4.6375573030
Y@Xmin:	3.2508297340	Y@Xmax:	5.3588364800	Y@Xrange:	2.1080067460

Date	Time	File Source
Feb 15, 2014	7:48:06 PM	CLIPBRD.PRN

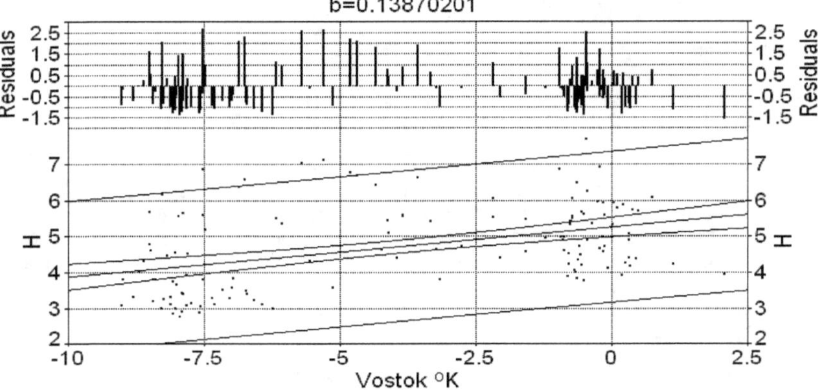

Figure 22B. Regression of H on Vostok temperature differences in the Cryptophytes' group. Number of points 133. Each point represents a paleorelevé. *Regression line*: centre line within two 95% confidence belts, inner for regression, outer for prediction. *Residuals and derivatives:* as identified. Numeric table:

Rank 1 Eqn 8160 [Line Robust None, Gaussian Errors] y=a+bx

r^2 Coef Det	DF Adj r^2	Fit Std Err	F-value
0.1771621874	0.1645031441	1.0492784400	28.205128880

Parm	Value	Std Error	t-value	95% Confidence Limits		P>|t|
a	5.267174932	0.136641794	38.54731973	4.996864868	5.537484996	0.00000
b	0.138702005	0.026116724	5.310850109	0.087036896	0.190367115	0.00000

Area Xmin-Xmax	Area Precision		
53.012170851	0.0000000000		
Function min	X-Value	Function max	X-Value
4.0160852438	-9.019982709	5.5529010625	2.0600000000
1st Deriv min	X-Value	1st Deriv max	X-Value
0.1387020051	-8.611474967	0.1387020051	-7.811760815
2nd Deriv min	X-Value	2nd Deriv max	X-Value
0.0000000000	-8.596767655	1.766285e-09	-7.914377691

Area Xmin-Xmax	Area Precision		
53.012170851	0.0000000000		
Function min	X-Value	Function max	X-Value
4.0160852438	-9.019982709	5.5529010625	2.0600000000
1st Deriv min	X-Value	1st Deriv max	X-Value
0.1387020051	-8.611474967	0.1387020051	-7.811760815
2nd Deriv min	X-Value	2nd Deriv max	X-Value
0.0000000000	-8.596767655	1.766285e-09	-7.914377691

Procedure	Minimization	Iterations	
LevMarqdt	Least Squares	7	
r^2 Coef Det	DF Adj r^2	Fit Std Err	r^2 Attainable
0.1771621874	0.1645031441	1.0492784400	0.9466054875

Source	Sum of Squares	DF	Mean Square	F Statistic	P>F
Regr	31.053431	1	31.053431	28.2051	0.00000
Error	144.22907	131	1.1009852		
Total	175.2825	132			
Lack Fit	134.86994	119	1.1333609	1.45316	0.23989
Pure Err	9.3591235	12	0.77992696		

Description: Cryptophytes chronosere

X Variable: Vostok oK

Xmin:	-9.020000000	Xmax:	2.0600000000	Xrange:	11.080000000
Xmean:	-3.903458647	Xstd:	3.4969151005	Xmedian:	-3.850000000
X@Ymin:	-7.950000000	X@Ymax:	-0.480000000	X@Yrange:	7.4700000000

Y Variable: H

Ymin:	2.7905996430	Ymax:	7.7236320400	Yrange:	4.9330323970
Ymean:	4.7257573906	Ystd:	1.1523444408	Ymedian:	4.6375573030
Y@Xmin:	3.0994353850	Y@Xmax:	3.9703048670	Y@Xrange:	0.8708694820

Date	Time	File Source
Feb 15, 2014	7:54:04 PM	CLIPBRD.PRN

Analysis of the Cryptophytes' w_{ab} *chronosere*

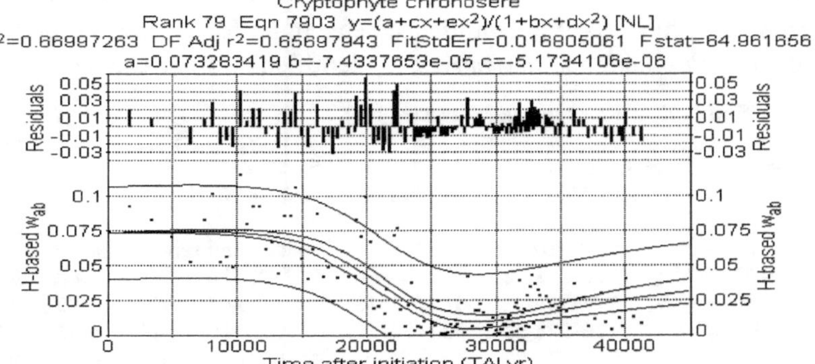

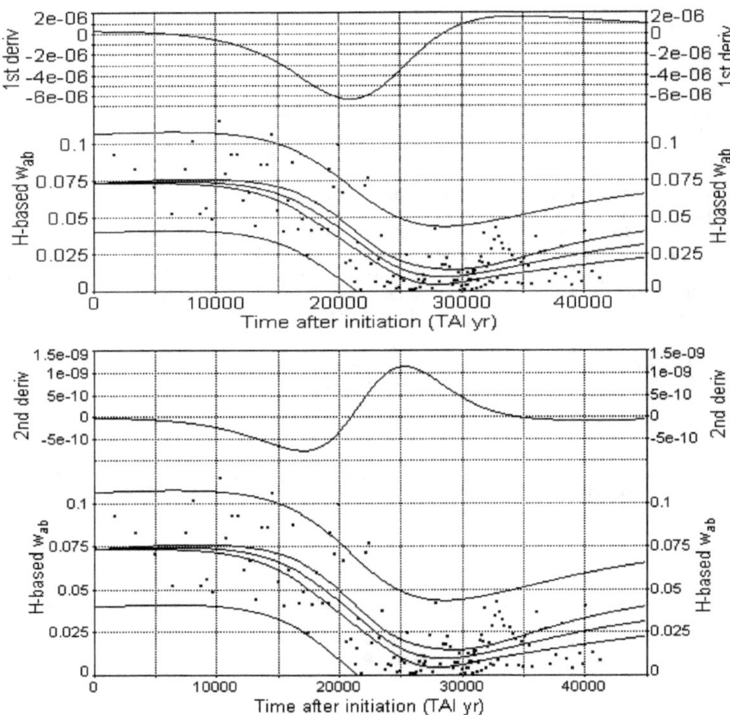

Figure 22C. Regression of H-based w_{ab} on time in the Cryptophytes' group. Number of points 133. Each point represents a paleorelevé. *Regression line*: centre line within two 95% confidence belts, inner for regression, outer for prediction. *Residuals and derivatives:* as identified. Numeric table:

Rank 79 Eqn 7903 $y=(a+cx+ex^2)/(1+bx+dx^2)$ [NL]

r^2 Coef Det	DF Adj r^2	Fit Std Err	F-value
0.6699726335	0.6569794301	0.0168050609	64.961656051

Parm	Value	Std Error	t-value	95% Confidence Limits		P>|t|
a	0.073283419	7.28072e-12	1.00654e+10	0.073283419	0.073283419	0.0000
b	-7.4338e-05	2.7163e-09	-27367.2539	-7.4343e-05	-7.4332e-05	0.0000
c	-5.1734e-06	9.76722e-08	-52.9670717	-5.3667e-06	-4.9801e-06	0.0000
d	1.64168e-09	6.94777e-11	23.62886988	1.50421e-09	1.77915e-09	0.0000
e	9.39353e-11	2.85437e-12	32.90934630	8.82874e-11	9.95831e-11	0.0000

Area Xmin-Xmax	Area Precision
1783.3007640	1.436691e-13

Function min	X-Value	Function max	X-Value
0.0099602778	28569.070141	0.0744252779	6357.9701627
1st Deriv min	X-Value	1st Deriv max	X-Value
-6.30805e-06	21038.469789	1.663179e-06	34660.618220
2nd Deriv min	X-Value	2nd Deriv max	X-Value
-7.68057e-10	17094.027016	1.150499e-09	25375.098127

Singularities [Data Range]
None
Singularities [All Other]
None

Soln Vector	Covar Matrix	SVD Cond
LvMrq/SVD	SVDecomp	1.672822e+20
r^2 Coef Det	DF Adj r^2	Fit Std Err
0.6699726335	0.6569794301	0.0168050609

Source	Sum of Squares	DF	Mean Square	F Statistic	P>F
Regr	0.073383303	4	0.018345826	64.9617	0.0000
Error	0.036148489	128	0.00028241007		
Total	0.10953179	132			

Description: Cryptophyte chronosere

X Variable: Tme after initiation (TAI yr)

Xmin:	0.0000000000	Xmax:	41138.000000	Xrange:	41138.000000
Xmean:	25563.609023	Xstd:	9209.0461001	Xmedian:	27522.000000
X@Ymin:	29851.000000	X@Ymax:	10198.000000	X@Yrange:	19653.000000

Y Variable: H-based wab

Ymin:	0.0008841120	Ymax:	0.1152327010	Yrange:	0.1143485890
Ymean:	0.0299449545	Ystd:	0.0288060116	Ymedian:	0.0191751790
Y@Xmin:	0.0744822050	Y@Xmax:	0.0093684580	Y@Xrange:	0.0651137470

Date	Time	File Source
Mar 16, 2014	6:01:58 PM	CLIPBRD.PRN

Cryptophytes chronosere
Rank 1 Eqn 8160 [Line Robust None, Gaussian Errors] y=a+bx
r^2=0.29805743 DF Adj r^2=0.28725832 FitStdErr=0.024226197 Fstat=55.624955
a=0.012390085

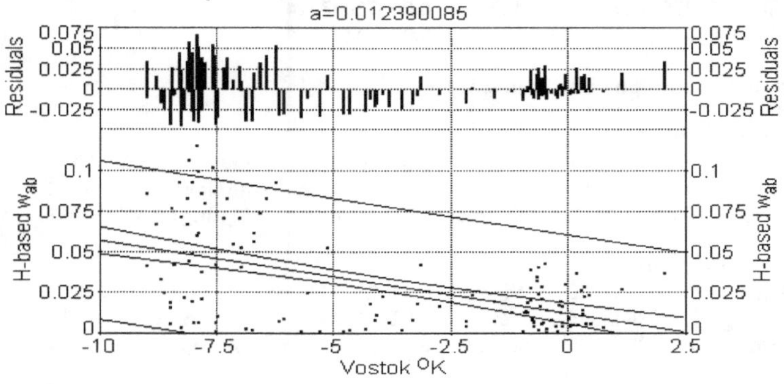

Figure 22D. Regression of H-based w_{ab} on Vostok temperature differences in the Cryptophytes' group. Number of points 133. Each point represents a paleorelevé. *Regression line*: centre line within two 95% confidence belts, inner for regression, outer for prediction. *Residuals:* as identified. Numerical table:

Rank 1 Eqn 8160 [Line Robust None, Gaussian Errors] y=a+bx

r^2 Coef Det	DF Adj r^2	Fit Std Err	F-value
0.2980574334	0.2872583170	0.0242261967	55.624955139

Parm	Value	Std Error	t-value	95% Confidence Limits		P>|t|
a	0.012390085	0.003154845	3.927319549	0.006149049	0.018631121	0.0001
b	-0.00449726	0.000602994	-7.45821394	-0.00569013	-0.00330439	0.0000

Area Xmin-Xmax	Area Precision
0.3106893021	0.0000000000

Function min	X-Value	Function max	X-Value
0.0031257288	2.0600000000	0.0529552949	-9.019982709
1st Deriv min	X-Value	1st Deriv max	X-Value
-0.004497260	-7.374155773	-0.004497260	-7.934676950
2nd Deriv min	X-Value	2nd Deriv max	X-Value
-3.44979e-12	-1.898840005	1.379909e-11	-8.401209501

Procedure	Minimization	Iterations	
LevMarqdt	Least Squares	7	
r^2 Coef Det	DF Adj r^2	Fit Std Err	r^2 Attainable
0.2980574334	0.2872583170	0.0242261967	0.9435936141

Source	Sum of Squares	DF	Mean Square	F Statistic	P>F
Regr	0.032646765	1	0.032646765	55.625	0.0000
Error	0.076885028	131	0.00058690861		
Total	0.10953179	132			
Lack Fit	0.070706735	119	0.00059417424	1.15406	0.4174
Pure Err	0.0061782926	12	0.00051485771		

Description: Cryptophytes vhronosere

X Variable: Vostok OK

Xmin:	-9.020000000	Xmax:	2.0600000000	Xrange:	11.080000000
Xmean:	-3.903458647	Xstd:	3.4969151005	Xmedian:	-3.850000000
X@Ymin:	-0.480000000	X@Ymax:	-7.950000000	X@Yrange:	7.4700000000

Y Variable: H-based wab

Ymin:	0.0008841120	Ymax:	0.1152327010	Yrange:	0.1143485890
Ymean:	0.0299449545	Ystd:	0.0288060116	Ymedian:	0.0191751790
Y@Xmin:	0.0860858430	Y@Xmax:	0.0370233820	Y@Xrange:	0.0490624610

Date	Time	File Source
Mar 16, 2014	6:12:49 PM	CLIPBRD.PRN

An anomaly

It is entirely possible that one or the other of the functional types will lack representation in some of the paleorelevés. The Phanerophytes have this defect in the Cwynar records. This is why the graphs drop to the zero level on both H and w_{ab} axes.[32] This anomaly has to be ignored, the data cannot be amended.

Functional types – a review

Chamaephytes

The one taxon potential energy H reaches maximum in the cold steppe stage of the climate cooling cycle (Figure 19B). Coincidental with this stage is a somewhat elevated level of instability (w_{ab} =0.05 or 10%) compared to the instability level (w_{ab}=0.01 or 2%) in the climate warming cycles (Figure 19C).

Hemicryptophytes

The one taxon potential energy H definitely identifies the Hemicryptophytes as functional type of the cold steppe climate (Figure 20B). The potential energy structure is substantially destabilised (w_{ab}=0.14 or 28%, Figure 20C) in the warming cycle.

Phanerophytes

The H parameter's tri-modality is most striking (Figure 21A). H takes a dive (Figure 21B) as the climate warming cycle takes effect. Recovery is noted in the Interglacial period. Instability

[32] For missing type T=0 and n=0, H is undefined, and w_{ab}= $2P_{ab}$=0.

reaches catastrophic level midrange during the intensive climate warming phase (w_{ab}=0.35 or 70%, Figure 21D).

Cryptophytes

This group comes into its own at the pea of the climate warming cycle. The peaking of Sphagnum is characteristic, indicating permafrost melt and extension of wetlands. We infer this from the high one taxon potential energy H (Figure 22A), and very high stability (0.5-w_{ab}=0.49 or 98%, Figure 22C) in the course of the climate warming.

Overview

Paradigma Novum

With this Book, the fourth on the topic written in succession, I completed laying the foundations for the holistic study of community energetics in my field. I used "Quantum ecology" to give the paradigm an identity. It is indeed, as Dieter Muller-Dombois, the master of Island Ecology characterised it, a new paradigm for the ecological sciences. Briefly:

1. In Quantum Ecology Max Planck's quantum theoretical principles join ecological theory.

2. Quantum Ecology revolves in its entirety about the potential energy structure's parameters, of which the most basic is the energy-based entropy,

$$E = nH = \ln C = \ln\frac{1}{P} = (T+n)\ln(T+n) - T\ln T - n\ln n$$

E as a value expresses in nats the potential energy level in T-totalled and n-valued complexes.

3. From E derives a set of axillary parameter equations including:

(i) $H = \frac{E}{n}$ -- one resonator's (average) energy level;

(ii) $w_{ab} = 1 - P_a^2 - P_b^2$ -- complexes' nH or H-based stability;

(iii) $dnH = nH(A+B) - nH(A) - nH(B)$ -- emergent potential energy in the complex, i.e. the size of the ghost energy cloud measured at the pivotal paleorelevé at which chronosere segments are joined;

(iv) $dH = \frac{dnH}{{}^{A}n}$ case G1 -- size of one resonator's ghost potential energy cloud when the pivot is considered a member of chronosere segment A;

(v) $dH = \frac{dnH}{{}^{B}n}$ case G2 -- size of one resonator's ghost potential energy cloud when the pivot is considered a member of chronosere segment B; and

(vi) $F = MA = \frac{T}{n}\frac{d^2y}{dt^2}$ and y=dnH -- potential energy based force.

Readers should see it clearly what shifting the study focus from calorific (trophic) flow to potential energy structure implies.

Operable equations

We were selecting shape functions and writing differential. It is emphasised that the function selected must have a credible shape. We declare credibility when the function captures what an expert will consider a natural trend upon which the trendless oscillations are superimposed. The first condition for a credible function's operability is its differentiability. The second is the credible function's ability to be parameterised by real data. Parameterisation is mechanistic in a high-level regression analysis of the target oscillogram. This is the step when we isolate the trend from random oscillations, find extreme points and draw diagrams of the function and the differential equations.

Energy footprints

I presented much detail on this topic in earlier essays. What is involved is an isolation of footprints, signals as it were, left in the complex's energy structure by the processes which govern the transitions. The most fundamental of these includes the historic phylogenetic effects and current environmental mediation. The present essay touched on the problem area of the latter, limited by the context imposed by the Cwynar chronosere's delimiters. The palynomorph taxonomy, the foremost among the delimiters, do not support the isolation of the phylogenetic footprint.

Further on holistic energetics

The potential energy model in Quantum Theory is holistic by the fact that $E=\ln P$ is defined for complexes of resonators. As it is illustrated in the Book, quantum analysis is focused on energy structures, structural oscillations, structural transitions, structural instability and structure based force. Structural transformations are targeted in time and space whenever the chronosere supports such a step. Energetics in the sense of calorific flow between parts of the complex does not enter the picture in quantum analysis. The analysis remains on the level of the resonator complex. This is not too difficult to understand if we recollect that P is a function of T and n, and not f_i and p_i or q_i. This is a good time for the student to review what he or she already read in the section concerning the unique properties of A. Rényi' entropy and information of order α, S. Kullback's discrimination information statistics, and L. Brillouin's information function, juxtaposed with the properties of Max Planck's energy-based entropy.

When quantum analysis?

Quantum analysis assumes that plant community's potential energy structure is a seamless fusion of the energy footprints of the basic processes which govern both directed and random compositional transitions in the perpetual assembly/disassembly dynamics of resonator complexes. This conceptualization of the community is not some rehashed analogue following the classical, often cumbersome graphical models, with no hope for operability.

Quantum analysis depicts reality in Nature as it presents itself on the resonator complex level such as the plant community, the metacommunity, the functional type, etc. This is baseline of generalizations, concerning the presentation of the resonator complex's governance rules to Science.

The nH vs. H dilemma

The value of nH is the potential energy level of the resonator complex. H is the energy level of one resonator. Being n neutral, H allows direct comparisons between cases. This leads us to what may be a revelation to the phytosociologist reader. Plant taxa are taken as objects, nothing more and nothing less than any physical resonator, which possess H amounts of potential energy on average.

Since the values of H and nH are the consequence of the plant traits actually measured, we shall see the community properties allowed us by the nats scale to see. This is not new, there is nothing in ecological studies that would not carry a scale effect.

Scale variables are everywhere and they define ecological perceptions. Therefore the first step in the presentation of results must be the clarification of scale for which the conclusions have their validity.

Whatever way we measure a trait, its magnitude to be achieved required energy to be put to work. The energy is the potential energy we gauge by nH or H. The unit of this energy is one nat. There is nothing new in this, considering that one nat is the unit on which diversity is measured traditionally.

Comparisons - a Gordian knot

The central theme of the Book is very much about comparison. But the compared events having to do with chronoseres may not be defined in universal time. This is definitely so with the pairing up Cwynar's carbon dated plant particle quantities with the Vostok temperature differences whose time scale is gas isotope based. This has only clinical solution since the time scales cannot be replaced. It amounts to shifting the focus from the individual pairs to arbitrarily delineated sections of the chronoseres.

So when we see an intensive peak on the Vostok graph and an explosive peak on the Cwynar parameters chronosere in its vicinity at an earlier time point, we know something is amiss with the time scales. The comparisons have to be broadened to general trends and reoccurring point patterns which we read from the shape function f(x), the residuals' oscillograms, and the derivatives.

Back to Cwynar's chronosere

The Late-Quaternary potential energy process of the vegetation is put on the analytical pallet in the Book in the eastern portion of Beringia adjacent to the western southern edge of the landmass actually affected by glaciation. The process is conceived as a continuous chronosere of energy state transitions of one resonator complex. The states are captured in the energy structure of paleorelevés at selected time points.

The plant particle chronosere in the Hanging Lake sediment is our source for evidence. Hanging Lake is in the eastern corner of Beringia close to the limits of the last continental ice sheath. Therefore the question of what this chronosere represents is legitimate. The plant particle composition is certainly not unique to Beringia. It is characteristic for the much broader region from which plant particles arrived. Beringian sources for wind-born plant particles may dominate the spectrum and in that way the results are biased for Beringia.

Ecological implications

The long-term vegetation process is a phenomenon of energy level transitions. This has a trended component with a trendless component superimposed. We measured these and found the trend sensitively linked with climate change.

Quantum analysis identified time segments and proxy temperature thresholds where directed transitions is a mandated conclusion, but the transitions differ in quality and quantity. These all depend considerably on plant taxon types and plant community type.

The inflexion region on the proxy temperature curve at which the climate cooling cycle ends and climate warming takes hold as the dominant trend is the period of highest stress. The mega sources of plant particles radically change. These are indications of biome level compositional transitions of the vegetation under sustained upward oscillations of temperature.

Proxy temperature thresholds, identified in the historic records, certainly offer themselves for direct prediction of vegetation energy states under the predicted global climate warming rates. In this regard we note that even the most conservative estimate of the current rate surpass by a wide margin the past rates. It is within the realm of real possibilities that sustained warming will create killing conditions for individual plants by corrupting its metabolism, lower its resistance and summary dieback on regional scales.

Which of the life forms are in greatest danger in the Arctic? Probably the Chamaephytes and Hemicryptophytes which have had their energy state peak during the climate cooling cycle. The destabilization of the cold step types' energy structure with the advent of climate warming is making this types' existence as a community assemblages very precarious under a new climate warming cycle.

The potential energy has bimodal oscillogram in the Phanerophytes' group. The two maxima brocket an important inflexion point on the Vostok temperature graph at the termination of the first phase in the climate warming cycle. The potential energy maximum of the Cryptogams is shifted well into the climate warming cycle. Coincidental with this is the return of the Taiga and Boreal forest vegetation into the once glaciated land within the source area of plant particles in the Hanging Lake sediments. This process coincidental with the seasonal melting of permafrost progressively deeper in the soil profile implies continued upward oscillations of the potential energy level of the Phanerophytes and Cryptophytes as climate warming progresses.

Bibliographic references

Borhidi, A. 1961. Klimadiagramme und klimazonale Karte Ungarns. Anna. Univ. Sci. Budapest, sect. Biol. 4: 21-50.

Brillouin, L. 1962. Science and Information Theory. 2nd ed. Academic Press, New York.

Çambel, A.B. 1993. Applied Chaos Theory: a Paradigm for Complexity. Academic Press, New York.

Cwynar, Les C. 1982. A Late-Quaternary Vegetation History from Hanging Lake, Northern Yukon. Ecological Monographs 52:1–24.

Cuffey, K M., Clow, G.D., Alley, R.B., Stuiver, M., Waddington, E.D., Saltus, R.W. 1995. Large Arctic Temperature Change at the Wisconsin-Holocene Glacial Transition. *Science*, 270(5235), 455-458,doi:10.1126/science.270.5235.455

Delcourt, P.A. and H.R. Delcourt. 1987. Long-term Forest Dynamics of the Temperate Zone. Ecological Studies 63, Springer–Verlag, New York.

Gleick, J. 1987. Chaos. Making a New Science. Penguin Books, New York.

Feynman, R. 1964. The Feynman Lectures on Physics. U.S.A. Addison Wesley. ISBN 0-201-02115-3. http://en.wikipedia.org/wiki/The_Feynman_Lectures_on_Physics

Hansson, M.E. 1994. The Renland ice core. A Northern Hemisphere record of aerosol composition over 120,000 years. Tellus 46(B), 390-418.

Hawking, S. (ed). 2011. The dreams that stuff is made of. Running Press, London.

Huang Xi and Zu Yuangang. 2001. The LES Population Model: Essential and Relationship to Lotka –Volterra Model. Ecological Modelling 143: 215-225.

Huang Xi. 2003. How do simple energy activities comprise complex behaviors of life systems? Ecological Modelling 165: 79-90.

Hulst, R. van. 2000. Vegetation dynamics and plant constraints: separating generalities and specific. Community Ecology 1:5-12.

Hultén, E. 1937. Outline of the history of arctic and boreal biota during the quaternary period: their evolution during and after the glacial period as indicated by the equiformal progressive areas of present plant species. Stockholm, Thule. 168 pp. Dissertation, Lund University.

Kerner von Marilaun, A. 1863. Das Pflanzenleben der Donauländer. Innsbruck, Wagner. – English rewrite: Conard, H.S. 1951. The Background of Plant Ecology. The Iowa State University Press, Ames.

Kitayama, K., D. Mueller-Dombois and P.M. Vitousek. 1995. Primary succession of Hawaiian montane rain forest on a chronosequence of eight lava flows. Journal of Vegetation Science 6: 211-222.

Kullback, S.M. 1968. Information theory and statistics. Dover, New York.

Lansner, T. 2008. CO2, temperature and ice ages. URL: http://icecap.us/images/uploads/CO2,Temperaturesandiceages-f.pdf

Magyari, E.K., Jakab, G., Sümegi, P. and Gy. Szöőr. 2008. Holocene vegetation dynamics in the Bereg Plain, NE Hungary – the Báb-tava pollen and plant macrofossil record. ACTA GGM DEBRECINA, Geology, Geomorphology, Physical Geography Series, Debrecen V. 3, 33–50.

Mason, J. 1990. The greenhouse effect and global warming. Information Office, British Coal, C.R.E. Stoke Orchard, Cheltenham, Gloucestershire, U.K. GL52 4RZ. -- To download copy, go to https://sites.google.com/site/statisticalecology/ then click on item 38 in Selected References section.

Milankovitch, Milutin. 1941. Canon of Insolation and the Ice Age Problem. Belgrade: Zavod za Udzbenike i Nastavna Sredstva. ISBN 86-17-06619-9. -- Theory summarised on Wikipedia http://en.wikipedia.org/wiki/Milankovitch_cycles .

Odum, H.T. 1971. Environment, power, and society. Wiley-Interscience, New York.

Orlóci, L. 1964. Global warming: the process and its anticipated phytoclimatic effects in temperate and cold zones. Coenoses 9: 69-74.

Orlóci, L. 2000. From Order to Causes. A personal view, concerning the principles of syndynamics. To download go to https://sites.google.com/site/statisticalecology/ then click item 14 in Selected References section.

Orlóci, L. 2006. Diversity partitions in 3-way sorting: functions, Venn diagram mappings, typical additive series, and examples. Community Ecology 7:253-259. --To download go to https://sites.google.com/site/statisticalecology/ then click item 9 in Selected References section.

Orlóci, L. 2008. Vegetation displacement issues and transition statistics in climate warming cycle. Community Ecology 9: 83-98. -- To download go to https://sites.google.com/site/statisticalecology/ then click item 7b in Selected References section.

Orlóci, L. 2012a. Statistical Ecology. The quantitative exploration of nature to reveal the unexpected. SCADA Publishing, Canada. Online Edition: https://createspace.com/3476529

Orlóci, L. 2012b. Self-organisation and Mediated Transience in Plant Communities. SCADA Publishing, Canada. Enlarged Online Edition: https://createspace.com/3585127

Orlóci, L. 2012c. Statistical multiscaling in dynamic ecology. Probing the long-term vegetation process for patterns of parameter oscillations. SCADA Publishing, Canada. Online Edition: https://createspace.com/3830594

Orlóci, L. 2013a. Quantum Ecology. Energy structure and its analysis. SCADA Publishing, Canada. Online Edition: https://createspace.com/4406077

Orlóci, L. 2013b. Quantum analysis of primary succession. The energy structure of a vegetation chronosere in Hawai'i Volcanoes National Park. SCADA Publishing, Canada. Online Edition: https://createspace.com/4452597

Orlóci, L. 2013c. On the Energy Structure of Natural vegetation. In search for community governance rules. SCADA Publishing, Canada. Enlarged Online Edition: https://createspace.com/4153484

Orlóci, L. and W. Stanek. 1980. Vegetation survey of the Alaska Highway, Yukon Territory: types and gradients. Vegetatio 41:

1-56. -- To download go to https://sites.google.com/site/statisticalecology/ then click item 41 in Selected References section.

Orlóci, L., Pillar, V.D. and M. Anand. 2006. Multiscale analysis palynological records: new possibilities. Community Ecology 7: 53-68.

Ortlóci, L. 2009. Multi-scale trajectory analysis: powerful conceptual tool for understanding ecological change. Frontiers of Biology in China 4: 158-179. -- To download go to https://sites.google.com/site/statisticalecology/ then click item 6b in Selected References section.

Petit, J.R., Jouzel, D. Raynaud, D., Barkov, N.I, Barnola, J.M., Basile, I., Bender, M., Chappellaz, J., Davis, J. , Delaygue, G., Delmotte, M., Kotlyakov, V.M., Legrand, M., Lipenkov, V., Lorius, C., Pepin, L., Ritz, C., Saltzmann, E., and M. Stievenard. 1999. Climate and atmospheric history of the past 420,000 years from the Vostok Ice Core, Antarctica. Nature 300: 429-436

Petit, J.R., Jouzel, J. Raynaud, D., Barkov, N.I, Barnola, J.M., Basile, I., Bender, M., Chappellaz, J., Davis, J. , Delaygue, G., Delmotte, M., Kotlyakov, V.M., Legrand, M., Lipenkov, V., Lorius, C., Pepin, L., Ritz, C., Saltzmann, E., and M. Stievenard. 2001. Vostok Ice Core Data for 420,000 years, IGBP PAGES/World Data Centre for Paleoclimatology Data Contribution Series #2001-076. NOAA/NGDC Paleoclimatology Program, Boulder CO, USA.

Planck, Max. 1901. On the law of distribution of energy in the normal spectrum. Annalen der Physik Vol. 4, p. 553 et seq. -- To download go to https://sites.google.com/site/statisticalecology/ then click item 40 in Selected References section.

Rényi, A. 1961. On measures of entropy and information. In: J. Neyman (ed.), Proceedings of the 4th Berkeley Symposium on Mathematical Statistics and Probability, pp. 547-561. University of California Press, Berkeley.

Vinther, B.M., H.B. Clausen, D.A. Fisher, R.M. Koerner, S.J. Johnsen, K.K. Andersen, D. Dahl-Jensen, S.O. Rasmussen, J.P. Steffensen, and A.M. Svensson. 2008. Synchronizing ice cores from the Renland and Agassiz ice caps to the Greenland Ice Core Chronology. J. Geophys. Res., 113, D08115, doi: 10.1029/2007JD009143

Schweingruber, F.H. 1996. Tree Rings and Environment Dendroecology. Paul Haupt, Stuttgart.

Schigolev, B.M. 1965. Mathematical analysis of observations. London Iliffe Books.

Shannon, C. E. 1948. A mathematical theory of communication. Bell System Tech. J. 27: 379-423.

Wildi, O. and M. Schüts. 2000. Reconstruction of a 405 yr. recovery process from pasture to forest. Community Ecology 1: 25-32.

Zólyomi, B., M. Kéri and F. Horváth. 1997. Spatial and temporal changes in the frequency of climate year types in the Carpathian Basin. Coenoses 12: 33-41.

Index

Appendix A

Episodes of glaciation

Each episode within the time frame of the Vostok chronosere has two dominant temperature cycles: a longer, sustained downward oscillation, followed by a shorter period of rapid upward oscillations. The top figure in Figure 4A shows this clearly for four complete episodes with 8 cycles. Each warming cycle ends in a short interglacial period. The current interglacial is exceptionally long.

The cycles follow a rather regular chronological schedule:

Cycle starts yr. BP	Cycle ends yr. BP	Cycle length yr	Type of climate cycle
410483	334101	76382	Cooling
334101	324129	9972	Warming
324129	265595	58534	Cooling
265595	238084	27511	Warming
238084	143790	94294	Cooling
143790	129324	14466	Warming
129324	24363	104961	Cooling
24363	8135	16228	Warming
8135	--	--	Current interglacial

The elapsed time between the successive interglacial episodes (distance between peaks) has increased.

The Vostok record set goes back a mere 422000 years. A longer version of the records exists which traces back the cycles to almost 800kyr.

The drivers

The Vostok series for CO2 and the temperature differences jaxtaposed in the next Figure 4C.

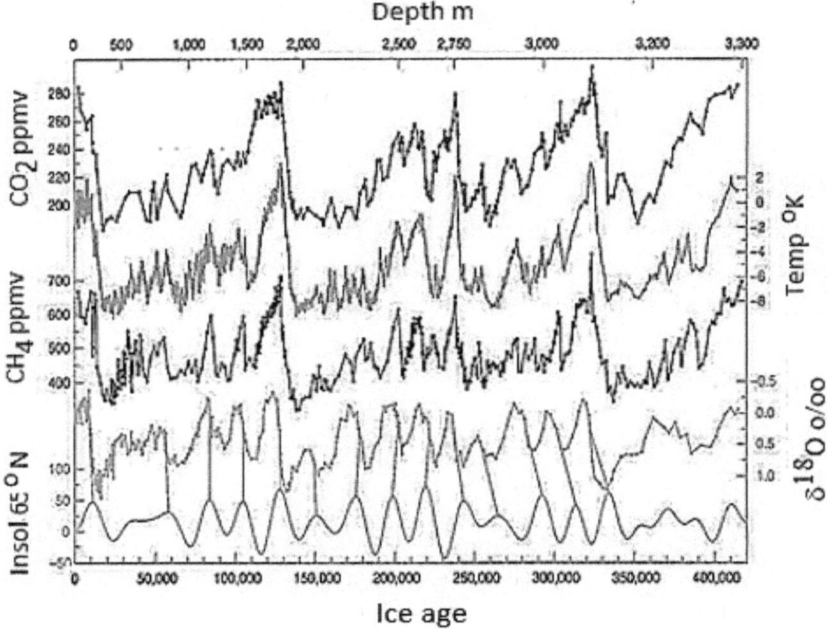

Figure App 1.. I borrowed this diagram from Wikipedia[33] to illustrate points I am about to make. The graphs span the entire temperature history of the ice core in the Vostok site. Not well seen at this magnification, the CO2 graph starts in year 2342 BP. Three other graphs are shown for the reader's perusal.

Observation of the concomitant oscillation in CO2 and temperature made me revisit specific conjectures regarding the debate under the umbrella title "drivers of the climate warming process", namely:

---

33 http://en.wikipedia.org/wiki/Milankovitch_cycles

1. CO2 driving temperature.
2. Temperature driving CO2.
3. Solar energy driving temperature change at critical points.

To me two facts are quite clear. The first one is that the rise of CO2 in the troposphere can be the cause of climate warming. Sir John Mason in his British Coal lecture of 1990 explained clearly the mechanism in CO2's climate warming effect and showed on pure, theoretical grounds that in 1990, CO2 already reached the critical volume in the atmosphere to cause substantial rise in the global average temperature. He made it clear, however, that climate warming would not be measurable as long as the oceans were still capable of absorbing heat. Once the oceanic thermal inertia is saturated, heat would be given up by the oceans and its climate warming effect would become observable. I believe that that time is right now, 24 years later, that the climate warming effect became obvious. But what means? Certainly not by averaging the available temperature records. The density of weather stations is totally insufficient for this in the most affected regions, namely the wide circumpolar belt of arctic tundra, taiga and boreal forest. I am also convinced that small, hardly measurable temperature rise can have mega effects.

I base my acceptance of global warming as an ongoing process in the troposphere is by its inevitability on theoretical grounds supported by real space/time manifestations such as the melting of the polar ice caps, disappearance of alpine glaciers, thawing of the permafrost in the Tundra deeper and deeper at an increasing rate over the years, migration of plant and animal species, and so forth. The warming process is in fact now instrumental in the strengthening of its main progenitor, the volume of CO2 and methane by being released in massive quantities in the thawing arctic region. This process is a synergetic *circulus vitiosus* which could have been triggered and certainly fed by human action but nothing in humanity's arsenal of remedies can stop it.[34]

Solar energy oscillations enter consideration as a cause (never an effect) of global climate warming different ways. One is predictable, linked to cyclic planetary motions (Milankovitch 1941). This component is responsible for driving the regular march of the climate warming and climate cooling cycles to alternating dominance. The other is the series of random events. The sunspots are behind this, which can elevate the total direct and indirect solar effect by causing the rising or falling volume of CO2 and other greenhouse gases in the atmosphere in unpredictable cycles.

I have my principal momentary interest in defining the effect of the warming and cooling climate cycles on compositional transitions in the vegetation cover and not in what causes these cycles to come and go. I measure the cycles' footprint in the potential energy structure of the vegetation. I this I rely on Max Planck principle: energy and entropy are interchangeable parameters of resonator complexes (complex systems). It is in fact in the generalised diversity theory where I place my present effort. My toughs along this line leads me to embed potential energy into my frame of reference as an indicator, a parameter to gauge the effect of climatic transience on the vegetation. That frame includes categories such as superposition, emergent sys-

[34]http://www.dailykos.com/story/2013/06/14/1216170/-Arctic-Methane-found-at-Amazing-Levels-by-NASA#

tem states, ghost systems, instability conditions and other measurable physical vegetation properties, such as probability and stochasticity.

Discouraging facts

Some interesting facts can be read from Figures 3A and B which are not at all encouraging regarding the future of the North and in deed for the entire Global environment. Consider these:

1. During the first 19k years, starting from the age of initiation, the bottom layer of the sediment core in Hanging Lake, the Vostok temperature oscillated downward by 2.20 ^{O}K units . This happens at a centennial rate of 0.012 ^{O}K. The corresponding slope is arc tan 0.012=0.69^{O}.

2. The centennial rate of warming during the next 14k years is 0.060 ^{O}K. The corresponding slope is arc tan 0.060=3.43^{O}. Under this rate of warming the elements of global pattern of the vegetation and indeed the entire biota was forced into major migration. This resulted in substantial shifts of the vegetation zones (Delcourt and Delcourt 1987, Orlóci 1994).

3. Knowing these facts we can contemplate a climate warming effect when sustained at a 3.5 ^{O}K per century rate, which appears to be the lower bound of the current estimates. This is 3.5/0.06 or 58 times the rate of historic warming. If such a rate of warming could dismantle and redesign the entire pattern of life on the global terra firma during the last warming cycle that ended about 6 millennia ago, we can ask what will a 3.5 ^{O}K centennial warming rate does to the Earth's biota. The answer I am afraid is a collapse, not excluding human civilization as we know it (see

Orlóci 1994).[35] As regards the North, I believe the minimum predicted rate of warming is sufficient to kill off individual plants and to lead to rapid, massive losses of species within decades. The disappearance of the tundra, broad extension of the dry steppe, and the disintegration of the boreal forest communities starting with a region a wide dieback is mandated.

[35]László Orlóci 1994 – download # 15 in selected references on page https://sites.google.com/site/statisticalecology/

Appendix B

Data set for Vostok and selected taxa

Entries in columns 5 to 11 are plant particle counts. Legend: Rel – Paleorelevé; TBP – time before present; TAI – time after initiation; °K – Vostok temperature differences; Pic – Picea; Aln – Alnus; Bet – Betula; Art – Artemisia; Sph – Sphagnum; Poa – Poaceae; Che/Am – Chenopodiaceae/Amaranthaceae.

Rel	TBP	TAI	°K	Pic	Aln	Bet	Art	Sph	Poa	Che/Am
1	0	41138	0	149	559	532	32	51	60	1
2	587	40551	-0.9	72	400	519	27	24	58	1
3	1148	39990	-0.65	22	95	149	5	7	17	0
4	1418	39720	-0.13	80	394	575	24	34	88	1
5	1683	39455	-0.88	25	111	216	8	16	48	0
6	2194	38944	-0.73	119	348	640	21	18	70	0
7	2680	38458	-0.02	38	135	234	14	11	27	1
8	3144	37994	-0.08	15	97	143	5	4	16	0
9	3584	37554	0.3	31	230	410	9	19	33	0
10	4003	37135	-0.73	89	384	825	16	55	77	0
11	4401	36737	1.13	30	225	397	4	16	27	0
12	4778	36360	0.25	38	172	268	4	128	25	0
13	5136	36002	0.18	34	103	208	8	13	21	0
14	5474	35664	0.2	118	667	999	21	48	98	0
15	5794	35344	-0.22	45	208	288	3	17	16	0
16	6096	35042	0.32	39	156	182	4	13	13	0
17	6241	34897	0.09	175	725	778	18	29	69	1
18	6381	34757	-1.59	37	213	237	4	15	19	0
19	6650	34488	0.37	159	666	949	14	27	72	1
20	6903	34235	-0.58	37	200	270	8	10	15	0
21	7141	33997	-0.63	40	146	210	4	6	9	0
22	7365	33773	-0.81	20	125	159	2	6	9	0
23	7575	33563	0.48	121	491	803	10	37	72	0
24	7773	33365	-0.57	26	145	147	1	3	7	0
25	7958	33180	-0.69	27	161	208	3	21	10	0
26	8132	33006	2.06	13	143	160	0	4	10	0
27	8295	32843	-0.78	86	524	1029	19	56	67	0
28	8447	32691	-0.52	12	166	252	3	12	15	0
29	8591	32547	-0.68	38	156	247	7	8	9	0
30	8659	32479	-0.56	158	895	1274	22	65	51	0
31	8726	32412	0.34	42	193	472	7	24	8	0
32	8852	32286	-0.53	80	279	1293	9	62	57	0

33	8972	32166	-0.88	19	33	383	2	18	18	0
34	9084	32054	-0.6	23	40	424	8	21	19	0
35	9191	31947	-0.13	24	25	471	4	17	14	0
36	9292	31846	0.03	64	87	1442	31	32	91	0
37	9389	31749	-0.82	12	13	329	9	14	16	0
38	9482	31656	-0.16	36	19	384	8	17	21	0
39	9572	31566	0.43	8	7	329	5	21	11	0
40	9660	31478	-0.46	66	26	986	13	39	54	1
41	9745	31393	-0.8	16	3	349	2	13	13	0
42	9830	31308	-0.74	76	10	1192	14	56	40	0
43	9914	31224	-0.37	31	2	356	5	17	13	0
44	9998	31140	-0.65	111	8	1448	21	78	71	1
45	10084	31054	-0.47	36	1	429	9	23	14	0
46	10171	30967	0.31	18	3	317	6	19	14	1
47	10260	30878	-0.18	11	1	292	4	18	8	0
48	10353	30785	-0.97	24	9	1430	19	69	56	1
49	10449	30689	-0.76	2	2	439	12	19	14	0
50	10550	30588	-0.79	1	1	355	3	10	5	0
51	10656	30482	-0.96	2	1	563	3	30	20	0
52	10768	30370	-0.26	7	4	2044	19	90	34	1
53	10886	30252	-0.15	1	0	356	2	5	5	0
54	11012	30126	-0.23	5	6	1472	34	41	58	1
55	11145	29993	0.74	2	3	345	6	5	12	0
56	11287	29851	-0.48	19	12	1552	77	12	72	3
57	11438	29700	-1.22	2	1	291	11	2	22	2
58	11600	29538	-1.59	1	0	255	8	0	14	2
59	11772	29366	-2.2	0	0	244	14	0	15	1
60	11955	29183	-2.19	4	6	810	71	3	107	8
61	12151	28987	-2.77	2	3	245	17	1	48	2
62	12359	28779	-3.96	1	1	203	23	0	41	2
63	12581	28557	-4.25	1	0	177	17	0	46	1
64	12817	28321	-3.24	5	0	440	64	0	133	2
65	13068	28070	-3.85	3	2	429	49	1	91	5
66	13334	27804	-3.17	0	1	213	28	0	44	2
67	13616	27522	-3.57	1	2	1048	54	1	120	3
68	13916	27222	-2.06	0	1	398	17	0	18	0
69	14233	26905	-3.34	0	0	242	9	1	23	0
70	14315	26823	-3.85	3	0	413	41	1	58	0
71	14399	26739	-4.11	1	1	222	22	1	47	2
72	14569	26569	-4.34	4	2	751	52	1	124	0
73	14744	26394	-4.14	2	0	284	16	0	54	0
74	14923	26215	-4.68	3	0	217	16	0	53	1
75	15108	26030	-5.32	5	9	150	16	0	34	1
76	15297	25841	-4.81	21	13	237	12	7	71	0
77	15492	25646	-5.71	8	8	89	14	2	64	2
78	15692	25446	-5.57	16	8	60	25	1	91	1
79	15897	25241	-6.19	23	10	66	28	1	87	0
80	16108	25030	-6.08	45	49	134	27	8	120	2
81	16324	24814	-6.88	5	1	30	5	2	29	0
82	16545	24593	-6.77	2	0	14	7	2	51	0
83	16772	24366	-7.49	11	6	26	9	1	51	4
84	17005	24133	-8.51	18	17	70	6	3	84	2

85	17244	23894	-7.53	53	60	171	19	8	77	2
86	17488	23650	-8.27	43	51	99	6	1	74	2
87	17739	23399	-7.84	49	46	75	11	2	87	3
88	17996	23142	-8.19	13	21	26	66	2	128	6
89	18258	22880	-7.89	19	28	45	28	2	67	4
90	18527	22611	-8.5	22	36	55	46	1	296	3
91	18802	22336	-8.46	37	106	239	57	12	181	4
92	19084	22054	-8.26	53	68	157	45	11	43	0
93	19372	21766	-7.54	43	32	61	73	4	109	4
94	19667	21471	-8.71	72	60	157	48	12	67	0
95	19817	21321	-8.52	120	99	221	102	17	123	7
96	19968	21170	-7.99	81	57	117	51	7	63	2
97	20276	20862	-7.82	100	138	257	90	18	125	4
98	20591	20547	-8.04	10	31	46	158	0	182	10
99	20913	20225	-8.12	16	24	40	59	1	97	2
100	21242	19896	-7.91	35	46	122	54	3	56	2
101	21578	19560	-7.36	90	169	286	104	19	150	10
102	21921	19217	-8.12	7	18	33	107	3	199	8
103	22095	19043	-8.41	13	14	26	198	6	224	7
104	22629	18509	-8.45	31	24	68	46	1	122	14
105	22994	18144	-7.92	12	8	32	110	2	256	7
106	23367	17771	-9.01	16	35	45	70	3	220	14
107	23748	17390	-8.63	46	38	70	54	6	153	19
108	24136	17002	-8.03	27	35	61	37	1	126	13
109	24531	16607	-7.53	36	43	79	63	8	183	9
110	24935	16203	-7.57	40	25	65	31	2	54	9
111	25347	15791	-7.89	25	31	107	43	7	101	5
112	25767	15371	-6.98	26	23	59	49	2	113	8
113	26194	14944	-7.16	9	8	26	65	3	143	16
114	26630	14508	-8.09	15	5	26	59	2	120	30
115	27075	14063	-9.02	24	5	33	74	1	93	28
116	27528	13610	-7.84	12	5	24	73	1	97	24
117	27989	13149	-8.09	7	13	24	128	0	91	25
118	28459	12679	-8.82	15	11	28	147	2	69	21
119	28937	12201	-6.71	9	5	27	123	7	94	30
120	29424	11714	-8.05	15	7	29	65	2	78	23
121	29921	11217	-8.3	15	10	28	96	4	105	24
122	30426	10712	-7.75	8	11	35	77	5	114	21
123	30940	10198	-7.95	16	8	25	94	4	106	19
124	31463	9675	-8.03	5	7	23	76	5	82	25
125	31995	9143	-6.74	13	5	16	94	1	59	34
126	32536	8602	-6.99	8	5	37	136	7	134	40
127	33087	8051	-7.6	10	11	24	89	3	132	22
128	33648	7490	-7.31	9	12	24	46	3	37	12
129	34797	6341	-5.14	19	18	46	49	2	79	11
130	36288	4850	-7.03	38	33	100	80	13	145	11
131	37841	3297	-6.45	32	30	98	79	13	105	8
132	39457	1681	-6.24	24	18	72	98	6	116	15
133	41138	0	-6.59	24	19	55	67	7	69	14

Data set for Vostok and plant functional types

TBP	Vostok	Ph n	Ph T	Ch n	Ch T	Hc n	Hc T	Cr n	Cr T
0	0	1	149	21	1411	12	82	3	233
587	-0.9	1	72	24	1218	13	76	3	152
1148	-0.65	2	23	18	341	7	27	2	34
1418	-0.13	1	80	25	1316	17	128	3	303
1683	-0.88	1	25	15	509	7	55	3	167
2194	-0.73	1	119	20	1340	10	114	3	485
2680	-0.02	1	38	14	519	8	42	3	215
3144	-0.08	2	16	15	305	4	19	3	69
3584	0.3	1	31	15	812	4	37	3	178
4003	-0.73	1	89	17	1569	11	102	3	295
4401	1.13	1	30	16	821	6	34	3	77
4778	0.25	1	38	17	1098	6	33	3	79
5136	0.18	1	34	14	394	2	22	4	75
5474	0.2	1	118	22	2339	7	114	3	401
5794	-0.22	1	45	15	682	8	24	3	121
6096	0.32	1	39	16	463	2	14	3	93
6241	0.09	1	175	21	1894	12	91	3	373
6381	-1.59	1	37	16	606	10	29	4	145
6650	0.37	1	159	21	2214	10	93	4	465
6903	-0.58	1	37	15	673	9	28	3	101
7141	-0.63	1	40	17	533	8	16	3	87
7365	-0.81	1	20	14	470	2	10	3	76
7575	0.48	2	122	22	1906	7	86	3	345
7773	-0.57	1	26	12	411	5	13	3	74
7958	-0.69	1	27	13	510	4	13	3	65
8132	2.06	1	13	11	384	6	21	3	57
8295	-0.78	2	87	23	2267	11	90	4	310
8447	-0.52	1	12	13	521	3	17	3	48
8591	-0.68	1	38	16	625	4	13	3	61
8659	-0.56	2	159	25	2632	7	82	3	331
8726	0.34	1	42	16	968	5	14	3	73
8852	-0.53	2	81	21	2264	10	71	3	311
8972	-0.88	1	19	16	583	3	21	3	113
9084	-0.6	1	23	15	630	4	22	2	85
9191	-0.13	1	24	16	693	2	15	3	128
9292	0.03	2	65	21	2122	9	113	3	429
9389	-0.82	1	12	13	476	3	18	4	71
9482	-0.16	2	37	16	574	7	27	3	109
9572	0.43	1	8	16	499	3	13	4	120
9660	-0.46	1	66	26	1909	10	65	3	584

9745	-0.8	1	16	18	727	6	18	3	104
9830	-0.74	1	76	23	1546	8	60	3	279
9914	-0.37	1	31	15	787	4	16	3	243
9998	-0.65	1	111	23	3009	13	87	3	749
10084	-0.47	1	36	16	798	6	22	4	201
10171	0.31	1	18	16	671	5	21	3	150
10260	-0.18	1	11	16	499	3	13	3	164
10353	-0.97	2	26	20	2598	12	78	3	1068
10449	-0.76	1	2	15	833	6	19	3	251
10550	-0.79	2	3	14	623	2	6	3	255
10656	-0.96	2	5	16	831	4	26	4	217
10768	-0.26	2	21	21	3040	12	53	5	733
10886	-0.15	2	4	11	505	1	5	3	435
11012	-0.23	2	20	21	2026	9	69	4	1515
11145	0.74	2	5	11	552	3	14	3	503
11287	-0.48	1	19	18	2940	13	96	3	2494
11438	-1.22	1	2	11	430	5	28	4	214
11600	-1.59	1	1	6	309	5	20	3	268
11772	-2.2	0	0	7	320	6	21	3	287
11955	-2.19	1	4	16	1209	16	149	3	476
12151	-2.77	2	3	9	359	5	58	3	125
12359	-3.96	1	1	9	291	8	53	3	92
12581	-4.25	1	1	7	317	7	58	3	117
12817	-3.24	1	5	10	647	14	164	3	117
13068	-3.85	1	3	13	619	15	127	3	294
13334	-3.17	0	0	12	335	10	57	3	49
13616	-3.57	1	1	14	1351	18	157	3	861
13916	-2.06	0	0	6	465	7	26	3	91
14233	-3.34	0	0	11	311	8	32	3	257
14315	-3.85	1	3	13	561	11	76	3	301
14399	-4.11	1	1	11	290	8	56	3	186
14569	-4.34	1	4	15	936	16	159	3	707
14744	-4.14	1	2	10	355	6	59	3	261
14923	-4.68	1	3	6	321	8	66	3	897
15108	-5.32	1	5	11	266	7	45	3	1419
15297	-4.81	1	21	13	410	9	91	3	1001
15492	-5.71	1	8	11	204	9	73	3	1282
15692	-5.57	1	16	13	149	12	111	3	84
15897	-6.19	1	23	14	167	9	102	3	281
16108	-6.08	1	45	15	354	8	157	3	237
16324	-6.88	1	5	10	206	6	40	3	660
16545	-6.77	1	2	8	171	7	67	3	812
16772	-7.49	1	11	12	175	9	73	3	204
17005	-8.51	1	18	16	319	13	132	3	329
17244	-7.53	1	53	17	417	15	112	3	304
17488	-8.27	1	43	16	407	13	110	3	536
17739	-7.84	1	49	14	283	15	117	3	55
17996	-8.19	1	13	13	248	13	194	2	64
18258	-7.89	1	19	14	185	11	103	2	212
18527	-8.5	1	22	14	231	13	346	4	150
18802	-8.46	1	37	15	502	14	232	3	26
19084	-8.26	1	53	15	322	6	60	3	28

19372	-7.54	1	43	11	211	9	130	2	718
19667	-8.71	1	72	17	340	13	100	2	42
19817	-8.52	1	120	15	547	13	164	2	90
19968	-7.99	1	81	15	301	12	92	2	194
20276	-7.82	1	100	19	604	13	170	3	102
20591	-8.04	1	10	11	260	13	243	2	71
20913	-8.12	1	16	10	149	12	130	2	20
21242	-7.91	1	35	14	255	9	88	2	13
21578	-7.36	1	90	18	645	12	207	4	38
21921	-8.12	1	7	12	183	15	261	2	16
22095	-8.41	1	13	12	274	17	316	3	49
22629	-8.45	1	31	14	172	11	166	2	33
22994	-7.92	1	12	13	185	13	337	3	34
23367	-9.01	2	17	14	197	13	291	3	50
23748	-8.63	1	46	16	211	10	219	1	28
24136	-8.03	1	27	10	155	10	174	2	28
24531	-7.53	1	36	13	220	10	246	2	33
24935	-7.57	1	40	9	143	10	126	2	15
25347	-7.89	1	25	13	210	8	120	2	22
25767	-6.98	1	26	11	145	9	156	1	17
26194	-7.16	1	9	8	123	10	221	3	37
26630	-8.09	1	15	6	95	8	216	3	18
27075	-9.02	1	24	13	130	7	172	3	23
27528	-7.84	1	12	7	109	11	161	3	23
27989	-8.09	1	7	11	189	9	163	2	31
28459	-8.82	1	15	11	204	8	143	2	20
28937	-6.71	1	9	8	174	8	181	3	33
29424	-8.05	1	15	8	113	8	151	3	21
29921	-8.3	1	15	11	152	7	181	3	21
30426	-7.75	1	8	9	148	7	186	3	25
30940	-7.95	2	17	8	140	7	182	2	11
31463	-8.03	1	5	9	121	8	167	2	28
31995	-6.74	1	13	7	126	8	153	3	36
32536	-6.99	1	8	7	193	9	251	2	26
33087	-7.6	1	10	10	147	7	214	3	19
33648	-7.31	1	9	8	90	7	90	1	8
34797	-5.14	1	19	9	122	9	133	1	13
36288	-7.03	1	38	14	252	7	207	2	19
37841	-6.45	1	32	11	236	8	167	2	16
39457	-6.24	1	24	13	212	9	179	2	14
41138	-6.59	1	24	11	159	8	127	1	9

Supplementary references

THE VEGETATION PROCESS: A holistic study of long-term community energetics in East Beringia

Authored by Dr Laszlo Orlóci

6" x 9" (15.24 x 22.86 cm)
Black & White on White paper
216 pages
ISBN-13: 978-1499142068 (CreateSpace-Assigned)
ISBN-10: 1499142064
BISAC: Science / Life Sciences / Ecology
ORDER FROM CREATESPACE E-STORE:
https://www.createspace.com/4760258

Process, as the Book uses this term, implies simultaneous execution of two fundamental functions in continuity. One creates complexity, the other reduces it. Ecologists refer to these as community assembly and disassembly. The process requires energy input which determines the momentary potential energy state of the community. This is measurably true in terms of Max Planck's energy-based entropy. We find potential energy increasing when new species (taxa, community elements) are added to or others proliferate in the community, and decreasing when species drop out or their performance declines.

Quantum analysis of primary succession: The energy structure of a vegetation chronosere in Hawaii Volcanoes National Park

Authored by Laszlo Orloci FRSC

List Price: **$30.00**

6" x 9" (15.24 x 22.86 cm)
Black & White on White paper
54 pages

ISBN-13: 978-1492788997 (CreateSpace-Assigned)
ISBN-10: 1492788996
BISAC: Science / Life Sciences / Ecology

The book revisits the classical idea that the potential energy structure of primary succession is a seamless fusion of foot-prints specific to basic processes which operate on all scales – phylogeny, environmental mediation, and chance. The idea is tested in quantum analysis of a vegetation chronosere in Hawai'i Volcanoes National Park. How is the test constructed? What are the outcomes? What do the results tell about primary succession not already known from other sources? Stated in the briefest of terms the test re-quires temporal species performance data.

ORDER FROM CREATESPACE ESTORE:
https://www.createspace.com/4452597

Statistical ecology

The quantitative exploration of Nature to reveal the unexpected

László Orlóci FRSC

List Price: $49.90

Add to Cart

Statistical ecology

Like 0

The quantitative exploration of Nature to reveal the unexpected

Authored by Laszlo Orlóci Ph.D.

The book's topics traverse many problem areas in univariate and multivariate data analysis, focussed on current ecological practice. The manner of presentation emphasizes reasoned methodological choices and encourages innovations consistent with the objectives, but mindful of the need to see clearly the regularity conditions which set limits for valid application of statistics in Ecology. The main text is accompanied by external appendices including a technical manual, 47 specialized application programs, and many data files taken from the exercises in the main text. For information please contact: lorloci@uwo.ca

About the author:
Orlóci is an INTECOL Distinguished Statistical Ecologist. He is external (academician) Member of the Hungarian Academy of Sciences, and regular (academician) Fellow of the Academy of Sciences of the Royal Society of Canada. He published over 100 papers in scientific journals, numerous monographs and books. His current essays on trajectory analysis, the rules of process governance, and the phylogenetic signal in vegetation transitions have considerable significance for evolutionary ecology and global change science. His present work on energy structures in metacommunities is seminal, pointing to a new direction.

Publication Date:	Aug 10 2010
ISBN/EAN13:	1453760520 / 9781453760529
Page Count:	372
Binding Type:	US Trade Paper
Trim Size:	6" x 9"
Language:	English
Color:	Black and White
Related Categories:	Science / Life Sciences / Ecology

List Price: $25.00

Add to Cart

Self-organization and mediated transience in plant communities

Like 0

What are the rules?

Authored by Dr. László Orlóci FRSC

A novel, signal theoretical solution is sketched out for the ecological problem of how to identify and quantitatively express the assembly rules of plant communities. A case study for testing the solution leads to the astonishing conclusion that the phylogenetic signal outperforms the current environmental signal in intensity close to 7 to 1. This indicates high stability and low inclination to environment mediated transience in the community.

About the author:
László Orlóci was born in Hungary in 1932. He holds degrees in forest engineering (DFE Sopron), forestry science and biology (BSF, MSc, PhD University of British Columbia), and DSc h.c. in biology (University of Trieste). Orlóci held appointments as NATO Science Fellow (University College of North Wales), professor (University of Western Ontario), and visiting professor at universities in the Americas, the Pacific, Asia, and Europe. He is an INTECOL Distinguished Statistical Ecologists, external (academician) member of the Hungarian Academy of Sciences, and regular Fellow of the Academy of Sciences of the Royal Society of Canada.

Publication Date:	Nov 11 2011
ISBN/EAN13:	1461028221 / 9781461028222
Page Count:	70
Binding Type:	US Trade Paper
Trim Size:	6" x 9"
Language:	English
Color:	Black and White
Related Categories:	Science / Life Sciences / Ecology

List Price: $30.00

Add to Cart

On the energy structure of natural vegetation

In search for community governance rules

Authored by Laszlo Orloci FRSC

Briefly about the book ...

Vegetation Science meets quantum theory in the energy-based entropy model of this book. The model is based on Max Planck's postulate that potential energy and entropy are interchangeable parameters in resonator complexes. What is a typical outcome of the model in vegetation studies? The model hands users a set of entropy estimates and probabilities based on which the strength and uniqueness of a metacommunity's energy structure can be characterised in comparative terms.

About the author:
Orlóci is an INTECOL Distinguished Statistical Ecologist. He is external (academician) Member of the Hungarian Academy of Sciences, and regular (academician) Fellow of the Academy of Sciences of the Royal Society of Canada. Orlóci published over 100 papers in scientific journals, numerous monographs, books and book chapters. His current essays on trajectory analysis, the rules of process governance, and the phylogenetic signal in vegetation transitions have considerable significance for evolutionary ecology and global change science. His present work on energy structures in metacommunities is seminal, pointing to a new direction.

Publication Date:	Jan 30 2013
ISBN/EAN13:	1482319373 / 9781482319378
Page Count:	46
Binding Type:	US Trade Paper
Trim Size:	6" x 9"
Language:	English
Color:	Black and White
Related Categories:	Science / Life Sciences / Ecology

List Price: $30.00

Add to Cart

Statistical Ecology. A reasoned approach.

Flexible computing in statistical ecology

Like 0

External appendix to accompany L. Orlóci's Statistical Ecology

Authored by Dr. László Orlóci

Problem flexible computing in statistical ecology

The Book describes more than 40 executable (.exe) computer programs and presents examples of application which correspond to the examples included in Statistical Ecology*. The programs are flexibly problem specific and conversational. They allow option-driven selective access to specific statistical tasks. Linkages are possible through standard output and input. The description includes in each case a brief introduction, a record of the start up dialogue, and detailed record input and output sets. The source code is in True Basic. The programs are compiled and linked on a 32 bit Windows XP system and tested up to Windows 7.
The executable program library, data files and a note to users are distributed free of charge to purchasers of the Technical Manual. Requests for download information should be directed to the URL address lorloci@uwo.ca. Prior to running the application programs, installation of a recent version of True Basic (see Internet for sources) on the user's system is strongly advised.
* Orlóci, L. 2010. Statistical Ecology. The quantitative exploration of nature to reveal the unexpected. Scada Publishing, Online Edition. Copies are available from the distributor https://www.createspace.com/3476529

Publication Date: Apr 05 2011
ISBN/EAN13: 1460972953 / 9781460972953
Page Count: 142
Binding Type: US Trade Paper
Trim Size: 6" x 9"
Language: English
Color: Black and White
Related Categories: Science / Life Sciences / Ecology

Bibliographic notes

- László Orlóci 2006

László Orlóci was born into a military family in Hungary in 1932. He holds degrees in forest engineering (DFE Sopron), forestry science and biology (BSF, MSc, PhD University of British Columbia), and DSc *h.c.* in biology (University of Trieste); held appointments as NATO Science Fellow (University College of North Wales), professor (University of Western Ontario), and visiting professor at universities in the Americas, the Pacific, Asia, and Europe. He is INTECOL's Distinguished Statistical Ecologists, external (academician) member of the Hungarian Academy of Sciences, and regular Fellow (academician) of the Academy of Sciences of the Royal Society of Canada.

Orlóci published over 100 papers in scientific journals, numerous monographs and books. His current essays on trajectory analysis, the rules of process governance, the phylogenetic signal in vegetation transitions, and the energy structure of the vegetation have considerable significance for evolutionary ecology and global change science.

Orlóci is married to Márta Mihály, a Sopron forest engineering alumna. Their daughter Martha Barbara is a Geography graduate of Western University, granddaughter Kathryn Orlóci-Goodison is enrolled in Forestry at Lake Head University in Thunder Bay, Ontario, and granddaughter Ruth Orlóci-Goodison attends high school in Cambridge, Ontario.

2014-04-04

About the Book

Process, as the Book uses this term, implies simultaneous execution of two fundamental functions in continuity. One creates complexity, the other reduces it. Ecologists refer to these as community assembly and disassembly. The process requires energy input which determines the momentary potential energy state of the community. This is measurably true in terms of Max Planck's *energy-based entropy.* We find potential energy increasing when new species (taxa, community elements) are added to or others proliferate in the community, and decreasing when species drop out or their performance declines. Data analysis which targets energy-based entropy is identified in the Book as *Quantum analysis.*

The Book's core objective is the probing of the historic vegetation process by quantum analytical techniques for the existence of regularities in climate driven potential energy oscillations. L.C. Cwynar's plant particle chronosere, from the Hanging Lake research site, supplies historic details. Temperature chronoseres from the Vostok and Renland polar ice cores allow the reconstruction of concomitant lower atmospheric temperature oscillations.

Quantum analysis reveals well-defined regularities. Theory and techniques are presented in detail. Numeric tables and graphs are included and explained. These lay solid foundation for self-study and interpretation. Many of the analytical tasks can be performed on a spreadsheet. Others require access to advanced software for calculus and regression analysis.

With such contents in the Book, it should find good use in the hands of researchers, and especially graduate students, planning projects in community energetics.

www.ingramcontent.com/pod-product-compliance
Lightning Source LLC
Chambersburg PA
CBHW051644170526
45167CB00001B/319

* 9 7 8 1 4 9 9 1 4 2 0 6 8 *